NEWTON · FARADAY · EINSTEIN

From Classical Physics to Modern Physics

NEWTON · FARADAY · EINSTEIN

From Classical Physics to Modern Physics

Tadayoshi Shioyama

Kyoto Institute of Technology, Japan

World Scientific

NEW JERSEY · LONDON · SINGAPORE · BEIJING · SHANGHAI · HONG KONG · TAIPEI · CHENNAI · TOKYO

Published by

World Scientific Publishing Co. Pte. Ltd.

5 Toh Tuck Link, Singapore 596224

USA office: 27 Warren Street, Suite 401-402, Hackensack, NJ 07601

UK office: 57 Shelton Street, Covent Garden, London WC2H 9HE

Library of Congress Cataloging-in-Publication Data

Names: Shioyama, Tadayoshi, author.

Title: Newton, Faraday, Einstein : from classical physics to modern physics / Tadayoshi Shioyama.

Description: New Jersey : World Scientific, [2021] | Includes bibliographical references and indexes.

Identifiers: LCCN 2021012080 (print) | LCCN 2021012081 (ebook) |

    ISBN 9789811235672 (hardcover) | ISBN 9789811236242 (paperback) |

    ISBN 9789811235689 (ebook for institutions) | ISBN 9789811235696 (ebook for individuals)

Subjects: LCSH: Newton, Isaac, 1642–1727. | Faraday, Michael, 1791–1867. |

    Einstein, Albert, 1879–1955. | Physicists--Biography. | Physics--History.

Classification: LCC QC15 .S447 2021 (print) | LCC QC15 (ebook) | DDC 530.092/2 [B]--dc23

LC record available at https://lccn.loc.gov/2021012080

LC ebook record available at https://lccn.loc.gov/2021012081

British Library Cataloguing-in-Publication Data

A catalogue record for this book is available from the British Library.

For any available supplementary material, please visit
https://www.worldscientific.com/worldscibooks/10.1142/12243#t=suppl

Preface

Our current lives are a result of scientific evolution, so awareness of the progress in scientific advances opens the doors to a new era. This book describes the lives of three great scientists who made remarkable discoveries—Newton, Faraday and Einstein. By focusing on their stories, the readers will understand that the common thread shared by them in their scientific journeys was a genuine enthusiasm to scholarship rather than any lust for fame or a sense of rivalry.

The progress of physics is surveyed as follows: Why is it that celestial bodies in the universe move according to Kepler's laws? This problem was elucidated by Newton who explained the motion of celestial bodies with three laws of motion as described in "Principia". Using Newton's principles, the heliocentlic theory, proposed by Copernicus and supported by Galileo and Kepler, was shown to be correct. The mechanics as outlined by Newton was called Newtonian mechanics that became dominant until the end of the nineteenth century.

Electromagnetic phenomenon was researched by Faraday who discovered the electromagnetic induction that transforms magnetism to electricity, Faraday's effect (magneto-optical effect) and so forth. Furthermore, based on such experimental results, Maxwell laid the foundation of electromagnetic theory. Newtonian mechanics and electromagnetic theory together constituted the two greatest theories in classical physics until the end of the nineteenth century.

However, from the end of the nineteenth century to the beginning of the twentieth century, some experimental results could not be elucidated by classical physics. The experimental results on the relation between intensity and frequency of blackbody radiation was difficult to be resolved by classical physics. In 1900, Planck derived Planck's formula which strictly adhered to the experimental results of blackbody radiation. From physical interpretation of the formula, Planck discovered the concept of "quantum". In 1905, introducing the concept of quantum, Einstein succeeded in theoretically elucidating the phenomenon of photoelectric effect where an electron was emitted from the metal irradiated with light. This successfully confirmed the concept of quantum, after which quantum mechanics was founded as an epoch-making mechanics following Newtonian mechanics.

According to relativity principle, the velocity of light measured in any coordinate system moving at a constant velocity (called inertial frame) should be constant always. Relativity principle was supported by Michelson–Morley experiment in 1887. However, in classical physics, the velocity of light was different in different inertial frames, and therefore, the experimental result by Michelson–Morley could not be explained by classical physics. In order to resolve this contradiction, Einstein revised the concept of time-space and founded relativity theory supporting relativity principle. Thus, Einstein contributed to the foundation of quantum mechanics and relativity theory which constitute the two greatest theories in modern physics.

These significant contributions to the progress of physics as mentioned above led to this book focusing on Newton, Faraday and Einstein. In a nutshell, Newton and Faraday contributed to the foundation of Newtonian mechanics and electromagnetic theory which comprise the two significant theories in classical physics, respectively. And, Einstein contributed to the foundation of quantum mechanics and relativity theory recognized as the two greatest theories in modern physics.

This book also describes other geniuses related to the three great Scientists—Galileo who was of considerable influence on Newton; Kepler who discovered laws of orbital motion of planets; Euler and Lagrange who founded the analytical mechanics which was a compromise between Newtonian mechanics and quantum mechanics; Volta who invented electric battery which played an important role in experiments by Faraday; Maxwell who succeeded in theorizing electromagnetic phenomena discovered by Faraday; Planck who proposed the concept of quantum which led to quantum mechanics and influenced Einstein's research on the photoelectric effect; de Broglie who proposed the concept of the particle-wave duality of electron; and Schrodinger who was influenced by de Broglie's idea and founded quantum mechanics.

The contents of the book is a result of the lecture entitled "History of science and technology" taught by the author at Kyoto Institute of Technology for about 10 years until 2006.

It helps the readers understand the progress in physics from classical to modern. Scientific knowledge and academic terms that are useful for understanding this book have been clarified with suitable explanations and appendices. This feature of the book will set it apart from general biographies.

It is written with the hope that youth worldwide on whom depends the future, will develop interest in science. The readers are to hold on to their hopes as these great scientists had to overcome adversities.

Acknowledgments

The author would like to show his appreciation to Professor K. Matsuda at Faculty of Law in Rikkyo University for visiting

Faraday Museum in the Royal Institution of Great Britain and taking pictures used in this book.

The author also thanks Dr. K. K. Phua (the editor-in-chief), Lakshmi Narayanan (editor) and Dr. A. Nargiza (acquisition editor), for facilitating the publication of this book. He also appreciates other authors for the books referred to in this book.

Tadayoshi Shioyama
October 2020
In Kyoto

Contents

Chapter 1
Isaac Newton

Sir Isaac Newton explained the motion of celestial bodies with the help of the laws of motion he discovered. The Newtonian mechanics he formulated constituted the two greatest theories of classical physics, together with the electromagnetic theory, theorized by James Clerk Maxwell on the basis of experimental research on the electromagnetic phenomena by Michael Faraday.

1

1.1 Upbringing

Newton's birth

Newton was born on January 4, 1643 (on December 25, 1642, according to the Julian calendar), at a manor house in Woolsthorpe (Fig. 1.1), near Grantham, Lincolnshire, England, one year after Galileo Galilei, the greatest thinker up to Newton's time, passed away. His father, Isaac Newton, Senior, was a farmer who had inherited the manor and the manor house and was a lord with seignorial authority over a handful of tenant farmers. A lord of this sort was called a yeoman in those days. He passed away of a disease three months before the birth of Newton.

When Newton was three years old, his mother, Hannah, married the rector Barnabas Smith of North Witham, a neighboring village. Smith did not want that Newton be brought to the new home. So Newton was brought up by his grandmother Margery Ayscough.

Fig. 1.1 The manor house where Newton was born, in Woolsthorpe. (Photograph taken by the author in June 2016.)

While Newton was facing this personal upheaval in his young life, many historically relevant events were unfolding. Charles I was beheaded in 1649 and the Puritan Revolution started. In 1653, Oliver Cromwell became Lord Protector. The civil war between the Puritans and the Royalists continued until 1658. Even in the countryside, the Puritan soldiers pursued the Royalists, and the political situation was unstable.

Admission into King's School

Newton's stepfather passed away in 1653. Newton continued living with his family with his grandmother, mother, a step-brother, and two stepsisters. One year later, he entered King's School (founded in 1528), at Grantham, in Lincolnshire. This school was considered to help you prepare for the entrance examinations of the Universities of Oxford and Cambridge. Grantham was a market town of a few hundred families and a key point in Lincolnshire, an important distribution center of agricultural products.

Grantham was 7 miles away from Woolsthorpe—a distance much too great for anyone to walk to school every day. So Newton took up lodgings with the Clarks. William Clark was an apothecary. His wife was a friend of Newton's mother. Catherine Storer was her daughter from her previous marriage. Catherine was two years younger than Newton and a merry girl. She helped Newton, who came from the countryside, to come out of his shell. Later, they got engaged.

A fight

Newton was not interested in studying at school and was in the lowest grade. He had received his earliest knowledge of basic chemistry from Clark while living with the family. But an event occurred that would change his life. One day, Newton was kicked hard in the stomach by a classmate who was in a superior grade. After school, in the churchyard, Newton challenged this much larger boy to a fight. Although Newton was not as robust as his antagonist, he fought with a lot of spirit and resolution and kept fighting until the other boy gave up and surrendered. During their fight, the schoolmaster's son came upon them and told Newton that he must treat the other boy like a coward and rub his nose against the wall. So Newton pulled him along by the ear and thrust his face against the side of the church. Still not content with his victory, before leaving the bully to nurse his wounds, Newton declared he would not rest until he had beaten the boy academically (White, 1998, p. 22). After this episode, Newton's studies quickly improved, so much so that he rose to the topmost position in the class. He became so interested in learning that the schoolmaster was surprised.

Two years' leave of absence

Newton's mother became wealthy because of increasing income from the manor. She decided that she would entrust the management of the manor to her eldest son. For that, she needed to train Newton in farm work. So she applied for two years' leave of absence, starting from 1658, making Newton quit school, despite the fact that he was doing well academically and had the highest grade, and attempted to make Newton work on a farm during that time. However, Newton tended to be constantly lost in thought, neglecting farm work. Basically, he was unfit for farm work.

Newton's uncle William Ayscough was the rector of an Anglican church after he graduated from the University of Cambridge. He persuaded Isaac's mother to allow him to go back to Grantham in 1660.

In the meanwhile, restoration of the monarchy in England was marked by Charles II retaking the throne. The political situation became peaceful.

1.2 Admission into the University of Cambridge

Sizars

On June 5, 1661, Newton enrolled at Trinity College, University of Cambridge (Fig. 1.2). Almost all of the 40-odd students were youths in superior social positions. They had prepared in public schools for entrance into the University of Cambridge and were wealthy. Cambridge recognized students in four categories: fellow commoners, pensioners, sizars, and subsizars. Fellow commoners were privileged students (White, 1998, p. 96), were dressed in sophisticated gowns, and dined at high table (Gleick, 2003, p. 20). Pensioners paid tuition fees and boarding fees and aimed for the rector. Sizars were exempted from tuition fees and boarding fees (subsizars paid tuition fees) and paid their way by emptying the bedpans, cleaning the rooms, and running errands for the more privileged students (White, 1998, p. 46). Sizars ate other students' leftovers. Newton entered as a subsizar and later became a sizar.

Fig. 1.2 Trinity College, University of Cambridge.

Though Newton's mother was wealthy, she spent little money on his schooling because she hoped that he would end up managing the manor and was not interested in a scholarship. That is the reason Newton entered the university as a subsizar. Although Newton provided menial services for privileged students, he enrolled at the University of Cambridge as a sizar because he yearned for a scholarship.

As a sizar, he was often treated with contempt by some in a superior social position or ignored. This made him an introvert. At Trinity Hall, two students shared the room. His roommate was in a superior social position and had many friends. When the friends visited the room, the room was noisy and Newton could not concentrate on his study. During such times, he would sit in the courtyard, viewing the nocturnal sky and quietly contemplating. One day, another student facing a similar problem happened to be in the courtyard and proposed to negotiate with the university to allow them to share a room. The proposal was accepted, and Newton and the other student could study in peace.

Curriculum

When Newton entered the University of Cambridge, the curriculum of university focused on the Middle Ages. The contents were theology, the classics, law, and medicine, and theology and the classics were treated seriously. Both natural science and mathematics were not included in the curriculum. Newton studied mathematics by himself. The traditional backbone of the university was made of the old notions of Aristotle, and logic, ethics, and rhetoric were the basis of philosophy. However, the universities on the Continent paid attention to the radical ideas of Galileo Galilei, Rene Descartes (Fig. 1.3), and Johannes Kepler (Fig. 1.4). At the library of the college, Newton studied the ideas of Descartes, Galileo, and Kepler and studied in detail

Fig. 1.3 Rene Descartes (1596–1650). (Portrait made in 1649.)

Fig. 1.4 Johannes Kepler (1571–1630). (Portrait made in 1610.)

Descartes's philosophy. Adding onto Aristotle's famous words, Newton said, "Plato is my friend; Aristotle is my friend, but my greatest friend is truth" (Gleick, 2003, p. 26). This note forces us to imagine Newton's future.

1.3 Academic Development in the Continent Focusing on Astronomy

Academic development in the Continent, which constituted the background of Newton's discovery, is described next, with a focus on astronomy.

Johannes Kepler

In 1543, Nicolaus Copernicus (Fig. 1.5) published *De Revolutionibus Orbium Coelestium* [*On the Revolutions of the Heavenly Spheres*] (Copernicus & Yajima, 1953). In this book, he described the heliocentric theory, as per which, the Sun is set at the center of the universe and the Earth revolves around the Sun. Galileo and Kepler, who analyzed the data of astronomical surveys by Tycho Brahe (Fig. 1.6), supported him.

Brahe's contributions to astronomy are beyond measure. He brought a revolution in instruments for astronomical

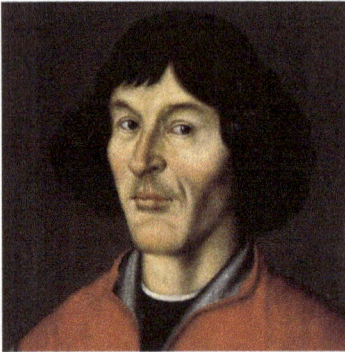

Fig. 1.5 Nicolaus Copernicus (1473–1543). (Portrait made in 1580.)

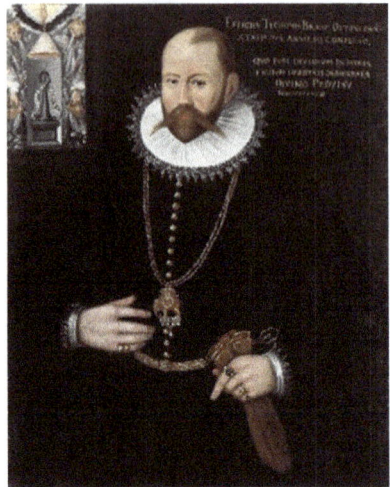

Fig. 1.6 Tycho Brahe (1546–1601). (Portrait made in 1596.)

surveys. Simultaneously, he introduced a revolutionary method in astronomical surveys. For example, when observing a planet, though until then planets were observed only at special times, he continuously observed the orbit of the planet and got a vast body of remarkably precise data of astronomical surveys.

When Newton began thinking about what controlled the orbital motion of celestial bodies, knowledge concerning how planets moved was the important basis on which the fundamental principle was researched. The laws of orbital motion of planets were discovered by Kepler, who analyzed the vast body of data of astronomical surveys by Brahe.

Galileo Galilei

Newton was greatly influenced by Galileo's work (Appendix 1.1) when he looked to understand physical phenomena. Newton generalized the basic thinking of the realm of motion by Galileo, unified the theories of Galileo and Kepler, and formulated Newtonian mechanics as the theory of motion.

The revolutionary points in Galileo's research work are as follows (Sugget & Oohasi, 1992):

- Time was introduced as an elementary quantity of a physical phenomenon.
- Natural phenomena that until then were expressed philosophically were expressed with quantitatively measurable quantities, such as weight and length.

Furthermore, on the basis of experiments, he explained the laws of physical phenomena in mathematical words for the first time. After him, this method of explanation became the most important tool of scientists.

—⋙⋘— ⋙ ⋘ —⋙⋘—

Explanation 1.1 Kepler's laws

■ Properties of an ellipse

To help you understand Kepler's laws, properties of ellipse are explained here. In Fig. E1.1, the semimajor axis a and the semiminor axis b are defined; two foci, f_1 and f_2, are defined; and the distance r between f_1 and point P—depicting a planet on the ellipse—is also defined. The distance r at the perihelion, where P is the nearest to f_1, is defined as r_1. The distance r at the aphelion is defined as r_2. The semimajor axis a is expressed by $(r_1 + r_2)/2$, and the semiminor axis b is expressed by $(r_1 \times r_2)^{1/2}$, where $(\ )^{1/2}$ expresses square root.

The semimajor axis was used in Kepler's third law. The perihelion appears Explanation 3.4 "The Perihelion Motion of Mercury," which is used to verify the correctness of the general relativistic theory.

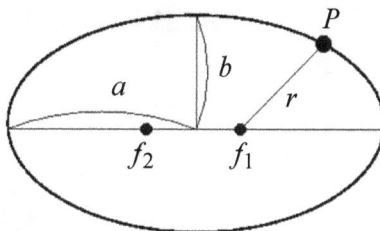

Fig. E1.1 Semimajor axis a and semiminor axis b: f_1 and f_2 are the foci.

■ **Kepler's first law**

Kepler analyzed an enormous amount of precise data on the orbital motion of planets by Brahe. Consequently, he found that Mars moves in an elliptic orbit around the Sun and one of the two foci is the placement of the Sun. This fact was found to be correct for all other planets. This is Kepler's first law (Fig. E1.2), according to which, a planet moves in an elliptical orbit, with the Sun as one of the two foci.

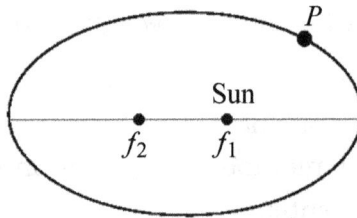

Fig. E1.2 Kepler's first law.

■ **Kepler's second law**

As per Kepler's second law (Fig. E1.3), the area swept by a line between a planet and the Sun is the same when the planet moves any distance on its elliptic path during the same interval. In 1609, this was described in *New Astronomy*, published in Heidelberg.

Afterward, Newton proved Kepler's second law was derived from the fact that the gravitational attraction between two point masses was inversely proportional to the square of the distance between them (this proportional relation was called the inverse square law). In other words, Kepler's second law helped to prove that if gravity obeyed the inverse square law, then a planet will move in an elliptic orbit.

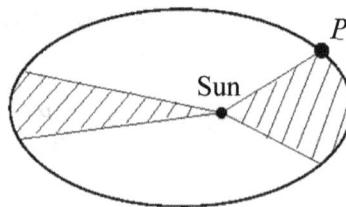

Fig. E1.3 Kepler's second law.

■ **Kepler's third law**

In 1619, with publication of his *The Harmony of the World*, Kepler gave his third law, in relation to two planets. According to Kepler's third law, the square of the ratio of an orbit's periodic times is equal to the cube of the ratio of its semimajor axis. Kepler's third law implies that the square of an orbit's periodic time is proportional to the cube of the semimajor axis.

Newton proved that if Kepler's third law holds true, then gravity on a planet will be inversely proportional to the square of its distance from the Sun. Kepler's third law helped to prove that if a planet moves in an elliptical orbit, then gravity obeys the inverse square law. To discover the three laws, Kepler analyzed the data of Mars using about one thousand papers for calculations.

—◊◊ ◊◊— ◊◊ ◊◊ —◊◊ ◊◊—

Appendix 1.1 Galileo Galilei

In 1564, Galileo (Fig. A1.1) was born in the neighborhood of Pisa, Italy, as the eldest among seven brothers. His father hoped for him to study medical science. In 1581, he made Galileo enroll at the University of Pisa. However, Galileo was interested in Euclidean geometry and Archimedes's method of applying mathematics to the problem of physics rather than in medical science. In 1585, he left the University of Pisa half-way, without earning his bachelor's degree in medicine.

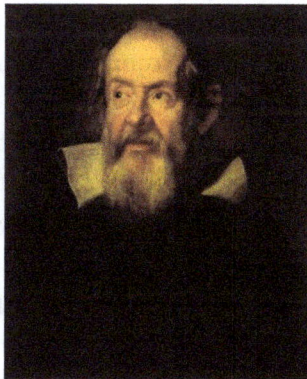

Fig. A1.1 Galileo Galilei (1564–1642).
(Portrait made in 1636.)

Afterward, in Firenze, Italy, Galileo began delivering private lessons on mathematics. He created a small balance and measured specific gravity. At the Siena School, he delivered lectures on mathematics. In 1587, in Firenze, while delivering private lessons, he began to write a manuscript on motion. In 1589, he was hired by the University of Pisa as a professor of mathematics. While researching the laws of motion, he reached the conclusion that a heavier body does not fall faster than a lighter body. In a cathedral, he measured the periodic time of a chandelier's swing by using his own pulse rate and found that the periodic time of the swing was always constant, independently of the amplitude of the swing. This led him to the principle of the pendulum clock (Explanation 1.2).

In 1592, at the University of Padua, Galileo joined as a professor of mathematics. In 1595, he gave his support to the heliocentric theory. In 1597, in a letter to Kepler, an astronomer and mathematician in Germany, he revealed that he supported Copernicus's theory. In 1602, he began research on pendulums and falling motion on slopes. Research was more convenient with slopes than with perpendicular falls for measuring the relation between time and falling distance, which was why he chose slopes to research falling motion. In 1604, he discovered the law of the pendulum and the law of falling bodies (Explanation 1.3).

In 1608, he discovered that in a free fall, velocity was proportional to time and the moving path of a cannonball was a parabola. At the beginning of 1609, Hans Lipperhey, an optician in the Netherlands, invented the telescope (Explanation 1.4). One story goes that when he happened to see the steeple of a church far away through a concave (凹) lens as the eyepiece and a convex (凸) lens as the object piece, to his surprise, the steeple appeared larger. This was a trigger for him to invent the telescope. Galileo, who heard the story, polished the surface of a lens to a desirable curvature and succeeded in making a telescope with a magnification of 30 with a combination of 凹 and 凸 lenses and presented it to the Governor of Venetia. On the other hand, Kepler fabricated a telescope with two 凸 lenses. This type of telescope widened the range of view but had the problem of chromatic and spherical aberrations.

In the summer of 1609, Galileo identified a crater on the surface of the Moon with his own telescope. The following year, he discovered four satellites of Jupiter and published *Sidereus Nuncius* [Report of Stars] and his name came to be known across Europe (Galilei *et al.*, 1976).

In 1632, Galileo published *Dialogo Sopra I due Massimi Sistemi del Mondo* [*The Dialogue Concerning the Two Chief World Systems*] (Galilei & Aoki, 1959–1961) and used the law of inertia to answer people who wanted to know why they did not feel the movement when the Earth rotated and revolved around the Sun. However, it was Newton, and not Galileo, who actually formulated the law of inertia.

In those days, any person who supported the heliocentric theory was treated as a heretic. *The Dialogue Concerning the Two Chief World Systems* was inspected by the Prosecutor's Office (the Inquisition Board). In 1633, Galileo was summoned to Rome and was brought to trial. *The Dialogue Concerning the Two Chief World Systems* was listed in Index Librorum. In 1638, *Discorsi e Dimonstrazioni Mathematiche Intorno a due Nouve Scienze, Attendi alla Meccanica ed ai Movimenti Locali* [*Dialogue Concerning Two New Sciences*] was published by Elsevier in the Netherlands because its publication was forbidden in Italy (Galilei, 1937–1948). In this book, Galileo explained the law of falling bodies with mathematical words and showed that the moving path of a cannonball was a parabola.

Thus, Galileo laid the foundation of modern science by discovering plenty of laws of physical phenomena.

— ⟫⟪ — ⟫ ⟪ — ⟫⟪ —

Explanation 1.2 Law of the pendulum

As shown in Fig. E1.4, when the weight set at end of the light, inelastic thread is swung in a vertical plane, the period of the pendulum is proportional to the square root of the thread's length. If the length of the thread is constant, then the period is constant independently of the amplitude of the swing. The law of the pendulum is applied to a pendulum clock and a metronome. You can adjust the period by changing the position of the weight.

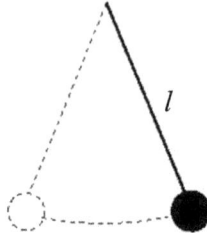

Fig. E1.4 A pendulum.

Explanation 1.3 Law of falling bodies

As shown in Fig. E1.5, when an object with zero velocity is released, the fall due to the Earth's gravity is called a "free fall." Then, the velocity of the falling body is proportional to the time from the time of release and the distance the body falls is proportional to the square of the time. The proportional coefficient contains only acceleration **g** of gravity and does not include the weight of the body. From this fact, it follows that when two bodies are released from the same height, they reach the ground at the same time, regardless of the weights of the bodies. In other words, the heavier body reaches the ground at the same time as the lighter body.

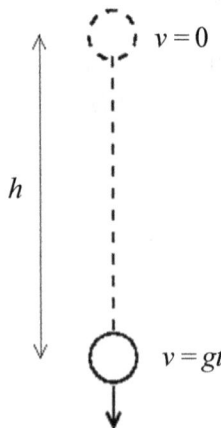

Fig. E1.5 A free-falling body.

1.4 Barrow, the Lucasian Professor of Mathematics

Descartes

The person in whom Newton was interested was Descartes. It was Descartes who applied algebra to geometry for the first time. In 1637, Descartes published *Discourse on the Method* (*Discours de la Methode pour Bien Conduire sa Raison et Chercher la Verite Dans les Sciences. Plus la Dioptrique, les Meteores et la Geometrie, Qui Sont des Essais de Cette Method*) at Leiden, the Netherlands. The book consists of three papers. The preface is called "Discourse on the Method" (Descartes, Miyake, & Koike, 1993), and the accompanying three papers are *La Dioptrique* (Descartes, Aoki, & Mizuno, 1993), *Les Meteores* (Descartes & Akagi, 1993), and *La Geometrie* (Descartes & Hara, 2013). Newton read a separate volume of *La Geometrie* translated into Latin with enthusiasm.

In 1663, Newton bought a book on astronomy. However, he could not understand the mathematics in the book and became aware of his lack of knowledge of geometry. He decided to read *Elements* by Euclid. He studied the geometry of Descartes and new algebra and analytical geometry. Studying mathematics, he devised his own proof method, different from the author's method.

Isaac Barrow

In 1663, when Newton began to study mathematics, Isaac Barrow (Figs. 1.7 and 1.8), who was a mathematician and a theologian, joined as the Lucasian Professor of Mathematics. He delivered lectures on natural philosophy (in those days, science was called thus) and optics. Newton attended Barrow's lecture. Barrow was a good teacher, who identified the genius of Newton and trained him.

Though Newton had no extra money because he was a sizar, he brought humble gifts for Catherine at Grantham, where he

Fig. 1.7 Isaac Barrow (1630–1677).

Fig. 1.8 Statue of Isaac Barrow. (Photograph taken by the author in June 2016 at Trinity Chapel.)

was lodged, and his stepbrother and stepsisters. He took care of his friends, brother, and sisters.

Barrow evaluated the creativity of Newton, who was twelve years younger than he and an innocent youth without a yen for fame. Barrow thought that he should look after Newton until he grew up because Newton showed all signs of becoming a great person someday. Receiving such respect from the professor made Newton study hard.

1.5 Creativity during the Plague

Plague

In 1665, Newton got a bachelor's degree in arts from the University of Cambridge. In summer that year, the plague spread quickly across London. The university was closed, and he went to his home in Woolsthorpe. The two years at his birth-place was the period when he showed the highest creativity in his life. He focused on three important problems concerning physics-astronomy, optics, and mathematics. Consequently, he made revolutionary discoveries and inventions in those realms.

Physics and astronomy

Newton first saw success in research on physics and astronomy. Twenty-four years before the birth of Newton, Kepler published the laws of orbital motion of planets in *New Astronomy* and *The Harmony of the World* (Kepler & Kishimoto, 2009) (Explanation 1.1) as an answer to the question on how planets moved. However, why planets should move following such laws was not explained.

One day, Newton was lost in thought in an orchard. The common belief that then an apple fell near him is due to biographer William Stukeley, who in the spring of 1726 visited Newton before Newton passed away. A free fall, expressed by the falling of a body, such as an apple, on to the ground was researched in detail by Galileo (Appendix 1.1; Explanation 1.3).

On observing the fall of the apple, Newton wondered whether the Moon is influenced by the same gravity as the one that influences an object like an apple on the Earth. He wondered why if the gravity of the Earth influences the Moon does the Moon not fall on to the ground the way an object like an apple does. Newton concluded that if the gravity of the Earth did not influence the Moon, by the law of inertia, the Moon would move linearly in the direction of its velocity and fly away into the universe. The fact that the Moon does not do so but revolves around the Earth shows that the Moon is actually constantly falling toward the Earth, from point a to point b, as shown in Fig. 1.9. He calculated the distance the Moon falls in 1 second and noted that down. For the following 20 years, this information was not disclosed (White, 1998, p. 92).

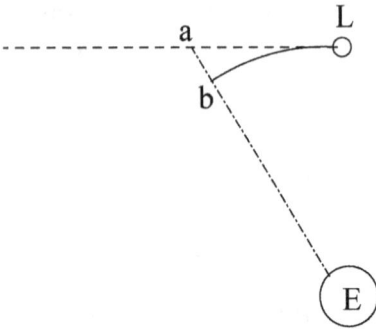

Fig. 1.9 The drop of the Moon (L) due to the gravity of the Earth (E).

Furthermore, Newton thought about what law should be satisfied by the gravity between the Sun and planets in order for the orbital motion of a planet to satisfy Kepler's third law, where the square of the periodic time of the planet's orbital motion was proportional to the cube of the semimajor axis of the ellipse with the Sun as a focus. Consequently, he discovered the inverse square law—the intensity of the gravitational pull between two point masses is inversely proportional to the square of the distance between them.

It is not as if all universal truths happened upon him as a divine message in a flash. Newton is supposed to have said, "I keep the subject of my inquiry constantly before me, and wait till the first dawning opens gradually, little by little, into a full and clear light."

Optics

During his research on optics, Newton performed an experiment using a triangle prism (Fig. 1.10). He allowed a sunbeam to pass through the prism. The ray of light was refracted and a beautiful rainbow of light emerged from the other side of the prism. He concluded that white light was actually composed

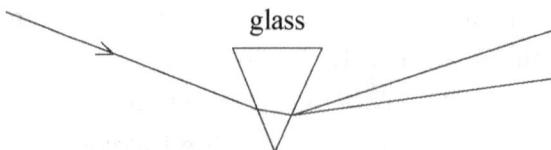

Fig. 1.10 Refraction of light using a glass prism.

of many different colors and light of each color had a different angle of refraction.

Mathematics

During his research on mathematics, Newton invented the methods for calculating differentiation and integration, influenced by Descartes's geometry and gradients and curves studied under Professor Barrow. Independently, Gottfried Wilhelm von Leibniz (Fig. 1.11) invented the methods of calculating differentiation and integration. However, Newton's invention was several years earlier than Leibniz's invention. Newton called his method the "method of fluxions." He thought that integral calculus was the inverse of differential calculus. Regarding differentiation as an elementary operation, he created the analytical method unifying different techniques such as area, tangential line, arclength of a curve, and maximum and minimum of function.

Fig. 1.11 Gottfried Wilhelm von Leibniz (1646–1716). (Portrait made in 1695.)

Reopening of university

Once the plague declined, in 1667, the University of Cambridge reopened. In March of that year, in the college chapel, Newton successfully cleared oral and written tests (White, 1998, p. 95). In October, he became a Minor Fellow of Trinity College. A Minor Fellow was provided a stipend and an allowance. It was most important that he could continue to research the previous subjects. He was given a room free of charge.

Though Newton was engaged to Catherine, the engagement was dissolved under mutual agreement because Catherine estimated that Newton would become a Minor Fellow and a new Minor Fellow was forbidden to marry for seven years. Newton did not forget her his entire life. Catherine married and got widowed twice, and all through her difficulties, Newton supported her financially.

— ·⟩⟩ ⟨⟨· — ·⟩⟩ ⟨⟨· — ·⟩⟩ ⟨⟨· —

Explanation 1.4 The principle of a telescope

According to Fig. E1.6, when a sufficiently distant object B is observed, the incident light is parallel and the image of B occurs at focal length f_o, in front of lens L_1. When B' is placed at focal length f_e, before lens L_2, the outgoing lights from B' are parallel and the image for B' is observed as a virtual image. Let the angle of incident light be θ, and let the angle of viewing the virtual image be θ'; then the magnification by the telescope is given as follows:

$$\text{Magnification} = \tan\theta'/\tan\theta = -f_o/f_e ,$$

where the negativity of the magnification means an inverted image.

The light passing through the part near the edge of the ⬚ lens, which has a remarkable curvature, bends sharply. However, the light passing near or through the center, where the lens curvature is the least, bends only slightly. The result is an unclear image. This problem is called a "spherical aberration." In Kepler's telescope, there was a problem of spherical aberration.

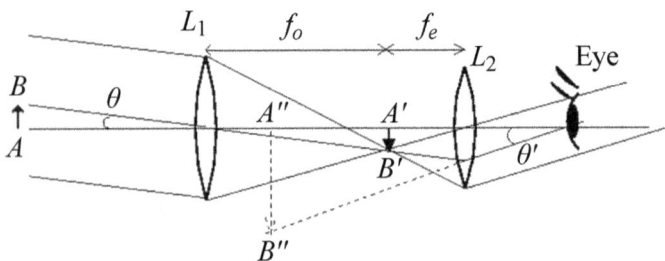

Fig. E1.6 How a telescope functions.

— ·⟩⟩ ⟨⟨· — ·⟩⟩ ⟨⟨· — ·⟩⟩ ⟨⟨· —

Newton went to Woolsthorpe to tell his family he had become a Minor Fellow. He stayed there, at his home, for a while. He talked about his telescope with his brother and sisters. When Galileo heard about the telescope invented by Lippershey in the Netherlands, he too decided to create a telescope. He used a 凸 lens as the object lens and a 凹 lens as the ocular lens and improved its magnification by polishing the lens to adjust the focal length. Kepler, on the other hand, increased the range of view using two 凸 lenses. However, a spherical aberration because the bending of the light was different at the center

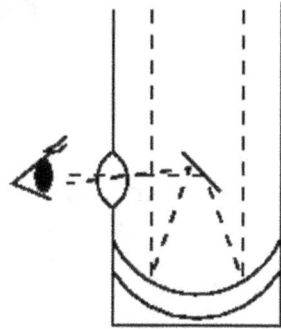

Fig. 1.12 Newton's reflecting telescope.

of the lens than at the edges of the lens led to a distorted image. To prevent this problem, Newton devised a different type of telescope. As shown in Fig. 1.12, he set a 凹 mirror at the bottom of the cylinder and in order to observe from the side of the cylinder, set a mirror using a 凸 lens as an ocular lens. The magnification of the reflecting telescope by Newton was 40 (Fig. 1.13). The huge reflecting telescopes set up in various astronomical observatories across the world today are a result of Newton's reflecting telescope.

In 1668, Newton came back to Cambridge from his home and got his master's degree in arts. In March, he became a Major Fellow and got an increment in his stipend and allowance.

1.6 Successor of Barrow

The Lucasian Professor of Mathematics

In 1669, his former teacher Barrow sent Newton's manuscripts to John Collins, who was the chief librarian at the Royal Society, in order to inform him about Newton's research works on mathematics. Collins immediately made the contents of the manuscripts known to noted mathematicians. Also, he sent the manuscripts to the president of the Royal Society, William Brouncker, after taking Newton's consent. However, later, Newton who had no ambition of standing out in the academic society and no lust for fame, asked for the return of his manuscripts. Consequently, the details of Newton's research works on mathematics remained unknown. However, most mathematicians understood the outline of Newton's research works through Collins.

In 1669, Barrow retired as the Lucasian Professor of Mathematics in order to devote himself to theology and nominated Newton as the successor.

In October 1669, Newton joined as the Lucasian Professor of Mathematics and delivered his first lecture in his new role in January 1670 (White, 1998, p. 163). Newton delivered a profound but difficult-to-understand lecture. Not a single student showed up for this lecture. Going forward, for the lectures that were especially difficult to understand, many students stayed away. But Newton did not lower the standard of his lectures for the students. He would go through the courtyard and reach the lecture room. If he did not find any student, he would wait for 15 minutes and deliver the lecture if some students arrived. If no student came, he would go back to his room. In his room, he would continue his research. The same story was repeated for almost every lecture for the next 17 years (White, 1998, p. 164).

As the Lucasian Professor of Mathematics, he first conducted research on optics. Since Aristotle, scientists had thought that solar light consisted of a single element. However, a chromatic aberration in the lens of his telescope led Newton to a different conclusion (Fig. 1.13). As mentioned above, Newton was aware that when white light passed through a prism, it split into lights of many colors, from red to violet. That is, Newton discovered that white light was composed of different colors and lights of different colors refracted at different angles.

In December 1671, Newton presented his own reflecting telescope to the Royal Society. His reflecting telescope with a high magnification got attention, and his name came to be known across London. In January 1672, after the presentation, he was elected as a Fellow of the Royal Society.

When Newton's reflecting telescope was demonstrated in the Royal Society, Robert Hooke, who was curator of experiments in the Royal Society, claimed that though he tried to fabricate a reflecting telescope, he could not complete it because of the plague. This was notified by Collins (White, 1998, p. 178). Afterward, Hooke criticized Newton's papers (as mentioned below).

Fig. 1.13 Newton's reflecting telescope. (Replica owned by the Royal Society, @Andrew Dunn; licensed under CC-BY SA 20, https://creativecommons.org/lisenses/by-sa/2.0.)

The first published paper

In 1672, Newton submitted his research result on the composition of white light to Henry Oldenburg, secretary of the Royal Society. "Theory of Light and Colors," published in the *Philosophical Transactions of the Royal Society*, was Newton's first published paper and got a favorable reception.

In this paper, Newton explained the experimental fact that white light was composed of plenty of elements and when a ray of light traveled through a prism, these were refracted at different angles, splitting the light into different colors. The prism did not add color to the light. He thought that light was a stream of particles and that a sound wave showed diffraction but light traveled along a straight line and so light was not a wave. However, Christiaan Huygens insisted that light was in fact a wave, and Hooke, who was in a position of refereeing the paper, criticized Newton's idea of light being corpuscular. Hooke insisted that color was added to the ray of light by the prism. Newton objected by saying that his paper was a scientific truth based on experimental facts and that they ignored the results of his experiments because his idea was contrary to theirs (Sootin & Watanabe, 1955). Newton's idea of light being corpuscular prevailed over others' ideas because of his good reputation, and its dominance continued till the idea of light being a wave was revived in the nineteenth century.

Newton regretted that he had published his paper because he felt it was a waste of time to argue with his critics about his findings. He desired neither reputation nor fame. He felt such desires led a person to waste precious time better spent on research and refection. He thought if he kept his presence unknown, he would have time for thought (Sootin & Watanabe, 1955).

In March 1673, he sent a letter informing Oldenburg of his resignation as Fellow of the Royal Society. But Oldenburg persuaded Newton not to resign.

The ideas of light being wavelike and light being corpuscular are both correct because the particle-wave duality of light (Appendix 1.2) has been verified.

In the latter half of the 1670s, many people close to Newton passed away. In 1677, Barrow and Oldenburg passed away. Afterward, his mother became critically ill and Newton went home to nurse her. He sat up entire nights consoling his mother, who was suffering from pain (White, 1998, p. 193). In spite of his devoted nursing, she passed away. Newton felt the sorrow and loneliness due to the loss of his mother for a long time.

─•≫❬❮•─ •≫ ❬❮• ─•≫❬❮•─

Appendix 1.2 The particle-wave duality of light

In the nineteenth century, Maxwell derived four electromagnetic equations and indicated that the electric field and the magnetic field satisfied the wave equation. He predicted theoretically the presence of an electromagnetic wave and predicted that light was identical to an electromagnetic wave. This prediction was afterward verified experimentally by Heinrich Rudolf Hertz. In experiments from 1886 to 1889, Hertz created electromagnetic waves with high-frequency electric vibrations and verified experimentally that in an electromagnetic wave, refraction, reflection, and polarization occurred as in light. He concluded that light was identical to an electromagnetic wave. That light was corpuscular was also proved a little later. In 1900, Max Karl Ernst Ludwig Planck, who pioneered quantum mechanics, proposed that the energy of light be expressed as the unit called "quantum" multiplied by an integer and that an energy quantum is proportional to the frequency of light. Light with the energy of a quantum was called a photon. In 1905, Albert Einstein proved using his theory on the photoelectric effect that light consisted of particles called photons (Explanation 1.5). Thus, the idea light being corpuscular was proved.

─•≫❬❮•─ •≫ ❬❮• ─•≫❬❮•─

Explanation 1.5 The photoelectric effect

As depicted in Fig. E1.7, when light is radiated onto a piece of metal (M), a photoelectron (e) is emitted. This phenomenon is called the photoelectric effect. Figure E1.7 shows the frequency v of light when such a phenomenon occurs. Planck proposed the energy quantum as hv (Appendix 3.1), where h is Planck's constant. Einstein called light with the energy of a quantum as a photon (Appendix 1.2).

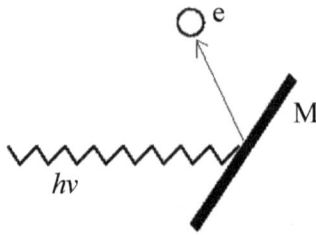

Fig. E1.7 The photoelectric effect.

The properties of the photoelectric effect are expressed as follows:

① If light with a wavelength longer than the limit wavelength λ_0 is radiated, then the photoelectric effect does not occur.

② The energy of an electron depends on the frequency of light and is independent of the amplitude of light.

③ The number of electrons is proportional to the intensity of the light radiated.

Explaining the photoelectric effect by the idea of light being a wave is difficult because:

• If light is wavelike, the photoelectric effect may be considered to occur by light radiating with a strong intensity. But this is contrary to property ① mentioned above.

• If light is wavelike, then an electron with more energy may be considered to be emitted for light radiated with more intensity. But this is contrary to property ② mentioned above.

In the beginning of the twentieth century, Einstein elucidated the photoelectric effect (Appendix 1.2 and Section 3.4 in Chapter 3).

—◊〉〈◊— ◊〉 〈◊— —◊〉〈◊—

Newton was unsatisfied with her death being recorded officially in the name of Barnabas Smith but was happy to see her buried alongside his natural father rather than his despised stepfather. After completing the management work for the inherited estate, he came back to Cambridge and keeping away from people, devoted himself to theology.

1.7 *Principia*

The three laws of motion

In 1687, Newton published *Philosophiae Naturalis Principia Mathematica* [*Mathematical Principle of Natural Philosophy*] (Newton & Nakano, 1977). This book is more commonly known as *Principia*. In this section, the process of the publication of *Principia* is described.

Principia is composed of an introduction and Books I, II, and III. Prior to 1666, Newton had an early idea of the three laws of motion (Explanation 1.6), which he described in the Introduction section: the first is the law of inertia (Fig. 1.14), the second is the law of motion (Fig. 1.15), and the third is the law of action and reaction.

As mentioned in Section 1.5, if the Moon did not feel the Earth's pull, then obeying the law of inertia, the Moon will move at a constant velocity in a linear direction and float away from the Earth. But this does not happen and the Moon revolves around the Earth because it feels the Earth's pull. Therefore, the law of inertia played an important role in his discovering universal gravitation (Explanation 1.6).

Galileo used the law of inertia to explain the heliocentric theory. When one person insisted on the geocentric theory and

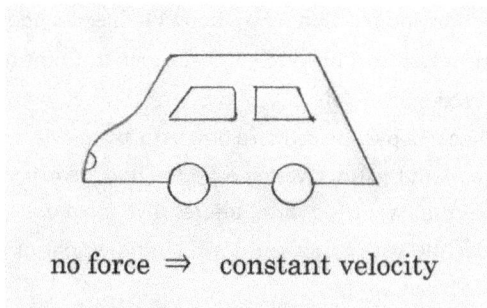

no force ⇒ constant velocity

Fig. 1.14 The law of inertia.

—•)}﴾﴿•— •)} ﴾﴿• —•)}﴾﴿• —

Explanation 1.6 The three laws of motion

■ **The law of inertia:** When a body is not subjected to a force or when two forces equilibrate, a body at rest remains at rest or one in motion remains at a constant velocity in a straight line.

For example, when brakes are applied to a moving train, the passengers inside it tend to fall in the direction the train was moving. This is because when brakes are applied, the train slows down but the passengers keep moving in the direction and at the velocity the train was moving because of the law of inertia.

Newton used the law of inertia to explain the motion of the Moon around the Earth. He concluded that the Earth exerted a force on the Moon—a gravitational pull—that ensured that the Moon revolved around the Earth rather than moving in a straight line. Thus the law of inertia played an important role in Newton's discovery of universal gravitation (Section 1.5).

■ **The law of motion:** When a force influences a body, the product of the resulting acceleration and the mass of the body is equal to the force.

In other words, as per the law of motion, force is the product of mass and acceleration. Acceleration is the differentiation of velocity and expresses the change of velocity per unit of time. According to Newton, when force is applied to a body, it produces acceleration and the acceleration is proportional to the force applied and inversely proportional to the mass of the body. The law of motion is the reason more force is required to move a heavier body (body with more mass) from a position of rest compared to a lighter body (body with less mass).

■ **The law of action and reaction:** When body 1 influences body 2 by force F, body 2 also influences body 1 by force $-F$, that is with the same magnitude but in the reverse direction.

For example, when a body is placed on a firm base, the gravity influencing the body equilibrates with the resistive force. So the body remains stable on the base. However, if the base is not firm, the resistive force does not influence the body because the base is pushed down by the weight of the body and so the body loses balance.

—•)}﴾﴿•— •)} ﴾﴿• —•)}﴾﴿•—

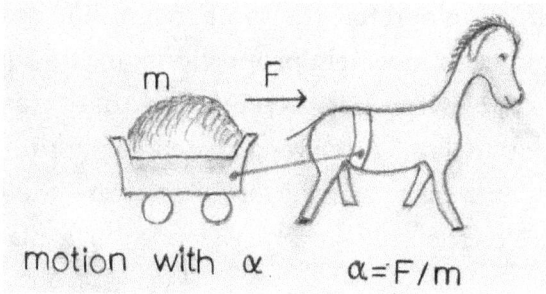

Fig. 1.15 The law of motion. F: force; m: mass; α: acceleration.

said that if the Earth rotated then if a body fell from above, it should fall toward the west, Galileo explained that a falling body felt gravity in a vertical direction but no force in a horizontal direction and because the body falls with a horizontal velocity due to the rotation of the Earth, following the law of inertia, the body will fall straight to the ground below and not toward the west.

Newton had already discovered the centrifugal force influencing a revolving body. The fact that the Moon revolves around the Earth indicates that the Moon has acceleration not because of a linear motion with a constant velocity but because of a force it faces due to the law of motion. This force is the centrifugal force and equilibrates with the gravity of the Earth, as shown in Fig. 1.18. The Moon does not fall onto the Earth because of the law of motion.

Centrifugal force (Fig. 1.18) is proportional to the product of the square of the angular velocity and the radius of the circle along which the body is moving (which in the case of the Earth and the Moon is the distance between them), and angular velocity is inversely proportional to periodic time. Consequently, the centrifugal force is inversely proportional to the square of the distance between the Earth and the Moon—due to Kepler's third law. The gravity equilibrating with the centrifugal force is inverse in direction and has the same magnitude as the centrifugal force and so is inversely proportional to the square of the distance. Thus, Newton (Fig. 1.17) inferred that "gravity obeyed the inverse square law." Afterward, instead of assuming that the Moon moves in a circular orbit, in which case the calculation was simple, Newton calculated the gravity if a planet has an elliptic orbit.

Halley's contributing to the publication of *Principia*

In January 1684, astronomer Edmond Halley (Fig. 1.16), who was interested in the orbital motion of planets, met Hooke at a London coffeehouse and had a chance to discuss the motion of celestial bodies. Halley asked Hooke whether he believed the force influencing the revolution of a planet around the Sun was inversely proportional to the square of the distance between them. Hooke said he did and, moreover, that he had the required mathematical proof (White, 1998, p. 190; Gleick, 2003, p. 124). But when Halley asked for the proof, Hooke failed to provide one. So Halley came to doubt Hooke's claim. In the summer of the same year, Halley visited Newton in Cambridge. He asked Newton what the orbital motion of a planet is if the Sun's gravity obeys the inverse square law. Newton immediately answered that it was elliptical. Newton said that he had made the calculations a long time ago and that though he could not give Halley the mathematical proof right then, he will send Halley the proof after redoing the calculations (White, 1998, p. 192). Newton, who had lost interest in mathematics and physics due to loneliness since Barrow had passed away, devoted himself to calculate mathematically the orbits of planets and then their gravitation using the clues from the mathematical calculations. In other words, Halley's visit to Cambridge encouraged Newton to take up mathematics and physics again.

Fig. 1.16 Edmond Halley (1656–1742). (Portrait made in 1690.)

Fig. 1.17 Statue of Isaac Newton. (Photograph taken by the author in June 2016 at Trinity Chapel.)

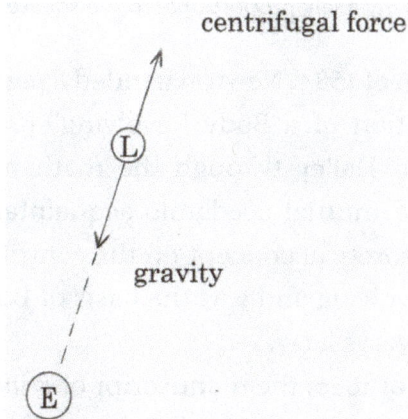

Fig. 1.18 Centrifugal force. E: Earth; L: Moon.

Newton proved mathematically that gravity is inversely proportional to the square of the distance, which resulted in Kepler's second law—a law concerning the area swept by a linear line between the Sun and a planet. Furthermore, Newton indicated that if a planet moved on an elliptical path under gravity, then the gravity was inversely proportional to the square of its distance from the Sun. In other words, if gravity obeys the inverse square law, then a planet will move on an elliptical path around an attracting body located at one focus of the ellipse. Inversely, if the orbit is an ellipse, then gravity obeys the inverse square law (Fig. 1.19). Newton completed a manuscript on the mathematical proof of the orbits of planets under his promise to Halley.

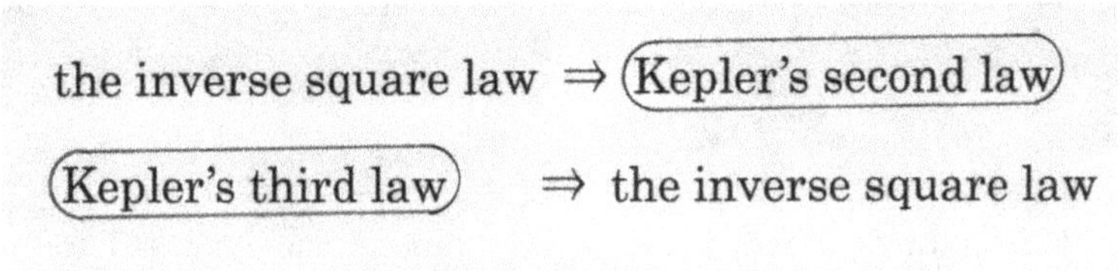

the inverse square law ⇒ (Kepler's second law)

(Kepler's third law) ⇒ the inverse square law

Fig. 1.19 The inverse square law and the elliptic orbit. According to the inverse square law, gravity is inversely proportional to the square of the distance.

In the autumn of 1684, Newton handed a paper of nine pages entitled "On Motion of a Body Revolving" [*De Motu Corporum in Gyrum*] over to Halley through the mathematician Edward Paget, who was a mutual academic acquaintance. This paper explained the theoretical concept on the centrifugal force influencing a body revolving and was the basis of Book I of *Principia*, published three years later.

In the spring of 1686, the manuscript of *Principia* was almost completed. Though it was not exactly perfect, as mentioned above, it was composed of an introduction and Books I, II, and III. Books I and II focused on force and motion. Book III described the application of the theoretical concepts described in Books I and II. Newton's three laws of motion were described in the introduction. The complete manuscript of Book III was brought to Halley on April, 4, 1687.

Principia (Fig. 1.20) was written in the form of propositions and deliberately made as difficult to understand as possible so that people with only a rudimentary knowledge of mathematics would not go about raising questions on it. That is, it was written in such a manner that you had to understand the previous propositions in order to understand the propositions given in the book.

With the help of Humphrey Newton, who served as Newton's assistant and transcribed his notes, Newton devoted himself to research and completing the manuscript of *Principia*, often forgetting to eat and rarely going to bed before 2 or 3 o'clock in the morning (White, 1998, p. 213).

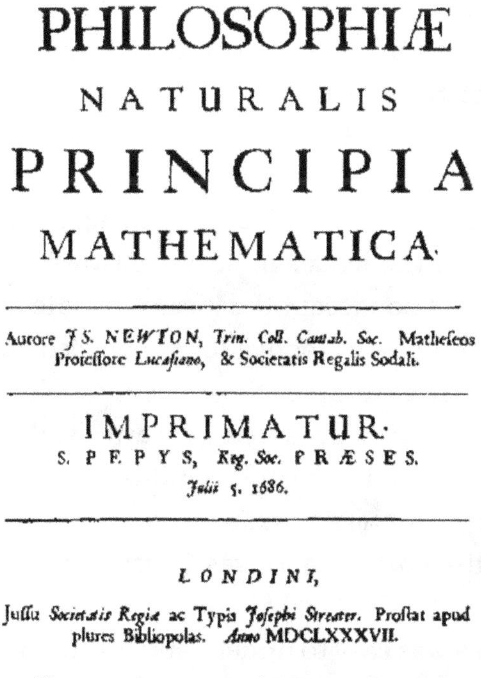

PHILOSOPHIÆ

NATURALIS

PRINCIPIA

MATHEMATICA.

Autore *JS. NEWTON*, Trin. Coll. Cantab. Soc. Mathefeos Profeffore Lucafiano, & Societatis Regalis Sodali.

IMPRIMATUR·
S. PEPYS, Reg. Soc. PRÆSES.

Julii 5. 1686.

LONDINI,

Juffu Societatis Regiæ ac Typis Jofephi Streater. Proftat apud plures Bibliopolas. Anno MDCLXXXVII.

Fig. 1.20 First-version cover of *Principia* (language: Latin).

In May 1685, before receiving the completed manuscript from Newton, Halley got approval for its publication during a meeting of the Royal Society. Because the Royal Society was bankrupt at the time, Halley used his own money to publish the manuscript. In July 1687, *Principia* was published and was evaluated as the greatest scientific book of the time. In this book, Newton indicated the principle that elucidated the motion of celestial bodies in the cosmos. The principle became a doctrine of the new epoch. He applied the principle to analyzing the motion of a body under the influence of gravity, that is, orbital motion, the motion of a thrown body, the motion of a pendulum, and the motion of a free-falling body.

Furthermore, he explained the law of universal gravitation, that is, any particle of matter in the universe attracts any other and the magnitude of attraction is proportional to the product of the masses of the two particles and inversely proportional to the square of the distance between them. The precession of the equinoxes of the Earth and the motion of the Moon receiving perturbation by the gravitation of the Sun were explained.

Remote action of force

The principle described in *Principia* was so high level that Newton was evaluated as the leader in science. However, scientists in the Continent did not accept the idea of remote action of force, which insists that the attractive force between two bodies can span large distances, that is, Newton's idea of universal gravitation. These scientists accepted Descartes's idea of proximity action of force, that is, the force influencing celestial bodies was transmitted in close proximity via aether, which filled all the space in the cosmos and was invisible and massless. The motions of celestial bodies were caused by vortices of aether. Descartes's idea was easy to understand for them. But this could not stop the worldwide praise for Newton's great achievement.

It was estimated that Newton was influenced by and enthusiastic about the idea of alchemy, that is, force is based on mysterious action. This was the reason why he rejected Descartes's idea.

1.8 Emergency at the University

Intervention of James II

After Charles II passed away of disease, the stubborn and unpopular James II succeeded to the throne. Because the king was Catholic, he thoughtlessly planned to force the entire country to become Catholic. But in those days, in Great Britain, there were both Catholics and Protestants living in harmony. The people did not support the king's plan.

The University of Cambridge was a Protestant fortress in those days. In February 1687, the king imposed unreasonable demands. James II ordered the university to install a Benedictine monk as a master of arts; the monk was exempted from the required examination and oaths to the Anglican church (Gleick, 2003, p. 146). A master of arts had the right to vote on the board of directors and had a voice in the management of the university. Therefore, when a person loyal to the king was installed as the master of arts, it was feared that the Crown will interfere in the university's functioning.

The university's firm attitude on Newton's advise

A devout Protestant, Newton suspected that the university was facing an emergency and advised the vice-chancellor of the university to stand firm against the tyranny of the king. The vice-chancellor created a document expressing firmness against the king's wishes, just as Newton had advised. The king lost his temper at the attitude of the university and ordered that eight representatives report to the Commission for Ecclesiastical Causes.

The board of directors elected Newton as one of the eight representatives. Before leaving for London, the representatives discussed the various measures. To avoid opposing the king, a compromise was proposed. Newton insisted that they remain firm in the face of the king's demands because if they were at the mercy of the king, then Catholics will prevail over Protestants.

At the Commission for Ecclesiastical Causes, the king's side exercised its authority but the university's side did not bow down to the king. The university's side successfully defended its autonomy and freedom of scholarship, foiling the thoughtless plan of the king.

At the University of Oxford, a Protestant fortress, defiant students rebelled against the king and the king sent equestrian soldiers. By this time, James II had lost support of the people.

In 1688, William III of Orange (Willem III van Oranje-Nassau), sovereign prince of Orange, the Netherlands, landed in and invaded England, supported by a group of British leaders. After a relatively peaceful revolt, he replaced James II as the next king of England. After being deposed, James II escaped to France.

In 1689, the University of Cambridge elected Newton (Fig. 1.21) as one of two members of Parliament. The reason was that Newton had shown a firm attitude against the thoughtless intervention of James II and had demonstrated leadership in helping the university face the king's aggression.

Fig. 1.21 Isaac Newton. (Portrait made in 1689.)

When as a member of Parliament Newton stayed in London, he saw the pleasant life the city had to offer, in contrast to the country, making him wish that he get a post in London.

1.9 Life in London

Exhaustion

In September 1693, Newton began suffering from lack of sleep and poor appetite, resulting from mental exhaustion due to the extraordinary concentration required in writing and the anxiety from the high expectations of the public after the success of *Principia*.

Newton (Fig. 1.22) thought that the vigorous life in London would offer succor against this exhaustion and the resulting fear. He expressed to a friend in London his desire to get a post in London. However, his friend could not help and Newton's exhaustion increased. To divert his attention from the exhaustion, he started carrying out research in an experimental room. It was here that he discovered the law of cooling,

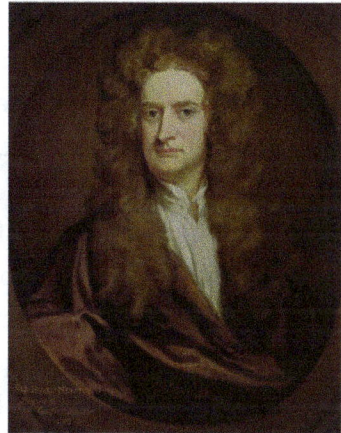

Fig. 1.22 Isaac Newton. (Portrait made in 1702.)

according to which, the thermal loss of a body is proportional to the difference in the temperatures of the body and its surrounding. Furthermore, he found that boiling and melting occurred at constant temperatures.

Master of the Royal Mint

Newton, who became increasingly exhausted due to sleeplessness, again sent a letter to his friend, stating his desire to get a post in London. In 1696, Newton got a post as Warden of the

Royal Mint. In 1699, he became Master of the Royal Mint. His work there involved recoinage.

He decided to reduce the number of silver coins in circulation. The government planned to cast new silver coins. However, when the new silver coins began circulating, people would store up the new silver coins thinking that they were higher in value than the older ones. As a result, only the old coins remained in circulation. So the recoinage did not produce the desired effect. In other words, as expressed in Gresham's law, "bad money drove out good money."

So Newton decided to scrap all old coins and replace them with new coins. To execute the plan, and prevent confusion, it was necessary to cast a vast amount of silver coins in as short a time as possible in the Royal Mint. One would estimate a civil servant to hard work. However, Newton decided to devote himself to the nation and be engaged in the difficult enterprise of recoinage. He successfully analyzed the work process and improved the efficiency of the process. He sought and caught counterfeiters and prosecuted them.

Until he resigned as professor at Cambridge in 1701, while concurrently working as Master of the Royal Mint, he worked with enthusiasm. He was satisfied with his life in London and his position as Master of the Royal Mint. He summoned his favorite niece to London. She was sociable and had a different personality from her uncle's. She was a witty, beautiful woman and took care of her uncle and played a role in his social circle.

Controversy with Leibniz

Though Newton lived a calm life, an unexpected controversy involving Leibniz occurred. Leibniz insisted that it was he who had invented differentiation and integration calculus, not Newton. Leibniz wrote a manuscript on differentiation and integration calculus in 1675 and got it published in 1686. But Newton had already invented differentiation and integration calculus in 1666, when the university was closed. Friends of Newton knew about his works on these topics. Today, it is thought that both scientists were stating the truth and had independently invented the calculus.

In November 1703, the Royal Society elected Newton as president for his great work. In February 1704, Newton presented his book *Optics* to the Royal Society (Newton & Shimao, 1983). It was not written in Latin but in English. Because of Hooke's criticism on the paper "Theory of Light and Colors" published in 1672, Newton did not publish the book immediately. After Hooke passed away in 1703, the book was published, 30 years after the paper had appeared. The book described refraction of light, reflection, rainbows, and the functions of mirrors and prisms on the basis of a series of Newton's experiments. In 1705, Newton, who had become famous worldwide as a scientist and the author of *Principia*, was knighted by Queen Anne. Newton was the first scientist given this honor.

Illness

At the age of 81, Newton was directed by his doctor to concentrate on his health when symptoms of kidney disease appeared. Afterward, pneumonia complicated the issue. His niece Catherine was taken aback by the seriousness of her uncle's illness and devoted herself to taking care of him. She decided to shift him away from the polluted air of Westminster in London, to Kensington.

Despite illness, Newton continued revising *Principia* and his research on theology. He attended meetings every week. On March 2, 1727, despite not feeling well, he went to attend a meeting of the Royal Society. He had pneumonia at the time, and attending the meeting put a great burden on his already deteriorating health. On returning home, he fell very ill and became bedridden. For two weeks, he alternated between coma and consciousness. During one of his conscious periods, he told John Conduitt, a colleague from the Royal Mint, and his nephew-in-law, with a smile that he had no intention of accepting the final ceremony as a Protestant (White, 1998, p. 360). On March 20, 1727, at the age of 84, the genius Newton passed away.

Newton enrolled at the University of Cambridge as a sizar. Afterward, he exhibited genius; accomplished great discoveries in physics, astronomy, and optics; and accomplished many inventions in mathematics. Through *Principia*, he explained the motion of celestial bodies. *Principia* became a doctrine of science in the new epoch. Newton, who noted down that "truth is my greater friend" in his youth, was buried as Sir Isaac Newton in Westminster Abbey. The inheritor constructed his monument in the corner called the "Scientist Corner," where other famous English scientists, such as Sir Charles Darwin and Maxwell were buried.

—◦❭❭ ❬❬◦— ◦❭❭ ❬❬◦ —◦❭❭ ❬❬◦—

Appendix 1.3 After Newtonian mechanics

Newton's laws of motion concern a dimensionless mass point. However, a real body has dimensions, and the motion of such a body should be derived from the motion of the mass point. The person who played an important role in identifying the motion of a body with volume was Jean le Rond d'Alembert (Yukawa & Tamura, 1955–1962). In 1743, he discovered the mechanics of a body with a volume without transformation, that is, the mechanics of a rigid body (Explanation 1.7).

New mathematical analysis for kinetic theory made progress, and rich mathematical methods contributed to the development of mechanics. Leonhard Euler and Joseph Louis Lagrange, great mathematicians of the eighteenth century, related the mechanics of a rigid body to Newtonian mechanics and formulated analytical mechanics, which was a mediator between Newtonian mechanics and quantum mechanics.

In the nineteenth century, electromagnetic phenomena were researched by Faraday, who proposed electric and magnetic fields, that is, force fields (Appendix 2.3). On the basis of Faraday's force field, Maxwell successfully derived electromagnetic equations consolidating electromagnetic phenomena. He predicted theoretically the existence of the electromagnetic wave and predicted that light was identical to an electromagnetic wave. Maxwell's prediction was verified by Hertz (Appendix 1.2). Afterward, the electromagnetic wave would indicate the application limit of Newtonian mechanics.

At the end of the nineteenth century, with the discovery of X-ray, cathode ray, and radioactive ray, the physical field proceeded toward the microscopic, involving atoms and such. In the microscopic field, assuming that Newtonian mechanics held true, statistic consideration by Maxwell and Boltzmann was useful in explaining the statistic thermodynamical properties.

However, the problem of black-body radiation could not be explained by classical physics until the end of the nineteenth century. This heat radiation involves an electromagnetic wave with a long wavelength and an invisible infrared light. The intensity of the heat radiation changes depending on the frequency, with the peak at some frequency. The peak shifts to a high frequency for a high temperature of the black body. which is on the basis of Wien's displacement law. Theoretical explanation was tried for the experimental data. However, the estimation by the classical theory did not coincide at all with the experimental data. Theoretical explanation failed to explain black-body radiation.

In 1900, Planck derived a theoretical formula strictly coinciding with the experimental data. This was Plank's formula (Appendix 3.2).

He was aware of the important quantity called the "energy quantum" (Appendix 3.1). The energy of light is given by the energy quantum as a unit multiplied by an integer. That is, he discovered that energy was a discrete quantity. Planck's discovery played the important role of starting quantum mechanics following Newtonian mechanics.

The special relativistic theory (Section 3.6, Chapter 3) was published by Einstein in 1905. The theory indicated the necessity of reconsidering the time-space concept. On the other hand, the progress of theoretical research on atomic structure had produced quantum mechanics (Section 3.13, Chapter 3). In the microscopic field, it was inevitable to generalize Newtonian mechanics fundamentally. Thus, at the beginning of the twentieth century, the application limit of Newtonian mechanics was indicated and Newtonian mechanics took its place as classical physics.

— ·≫ ≪· — ·≫ ≪· — ·≫ ≪· —

Explanation 1.7 Rigid-body mechanics

A rigid body is a system of mass points in which distances between the mass points do not vary (Fig. E1.8).

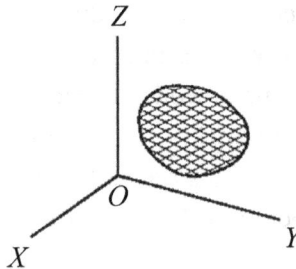

Fig. E1.8 A rigid body.

The state of the mass point is expressed by indicating the position in a 3D space. On the other hand, the state of the rigid body is expressed by indicating not only its position in a 3D space but also its rotational axis and the angle around the axis because of the body's volume.

The equation of motion is expressed as follows:

All mass × acceleration of center of gravity = sum of forces.   (E1.1)

Temporal differentiation of sum of angular momentum of
mass point = sum of moments of forces.   (E1.2)

Though in Newtonian mechanics, the equation of motion is expressed by indicating the equation of acceleration of the mass point, the equation of a rigid body's motion is expressed by indicating not only the equation of the center of gravity of the rigid body but also Eq. (E1.2), expressing the rotation of the rigid body because of its volume. Equations (E1.1) and (E1.2) decide the position and posture of the rigid body. Angular momentum is proportional to the product of velocity and the distance between mass point and rotation axis. The moment of force is proportional to the product of force and the distance from the rotation axis.

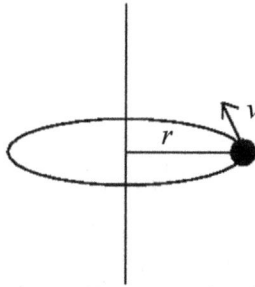

Fig. E1.9 Revolution around an axis.

For example, according to Fig. E1.9, when force is given to a mass point at position of radius r revolving with velocity v around the axis, Eq. (E1.2) indicates that the temporal change of the angular momentum is equal to the moment of force. When there is no moment of force, the right-hand side of Eq. (E1.2) is 0, and angular momentum, which is proportional to the product of radius and velocity, is constant. Therefore, in the case of absence of moment of force, as shown in Fig. E1.9, a decreasing radius increases the velocity v. Velocity is given as the product of radius and angular velocity, and this is the reason why in figure skating, when a person is spinning, bringing the spread arms close to the trunk, which acts as the axis, increases the person's angular velocity.

Explanation 1.8 Minimum-action principle

The minimum-action principle is a principle in physics using the variational method. We consider the moving path of the mass point from time t_1 to time

t_2 in a force field. We define Lagrangean L as the difference between kinetic energy and potential energy. For example, when a mass point is placed in a field of gravity on the ground, the mass point has a larger potential energy for the higher position. Kinetic energy is proportional to the square of velocity of a mass point. We define action integral as the integral of L from time t_1 to t_2. The minimum-action principle gives the equation deciding the moving path of the mass point with the minimum-action integral. This equation is called the Euler equation.

We define the path $q(t)$ as the position at time t between t_1 and t_2. In Fig. E1.10, it is assumed that a new path $q_\varepsilon(t)$ is given by the variation of path $\varepsilon\delta q(t)$. The variation is assumed to be zero at the end points $q(t_1)$ and $q(t_2)$. ε is assumed to be small.

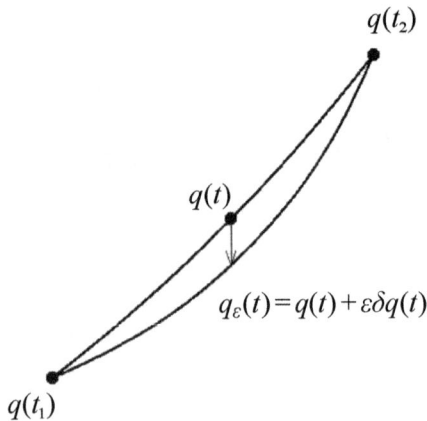

Fig. E1.10 Variation $\varepsilon\delta q(t)$ of the path.

The Euler equation is derived from the necessary condition that if the path $q(t)$ has the minimum action integral, the variation of the action integral should be zero for any variation in the path. The Euler equation decides that the mass point moves according to Newton's second law. That is, Newton's motion equation is expressed by using the minimum-action principle. Using the minimum-action principle, Euler and Lagrange expressed Newtonian mechanics in a mathematically excellent form (Yukawa & Tamura, 1955–1962).

References

Copernicus, N., & Yajima, Y. (trans. 1953), 天体の回転について. Iwanami Shoten, 岩波書店. (1543). *On the revolutions of the heavenly spheres* [*De revolutionibus orbium coelestium*].

Descartes, R., & Akagi, S. (trans. 1993), 気象学. Hakusuisha, 白水社. (1637). *Les meteores.*

Descartes, R., Aoki, Y., & Mizuno, K. (trans. 1993), 屈折光学. Hakusuisha, 白水社. (1637). *La dioptrique.*

Descartes, R., & Hara, R. (trans. 2013), 幾何学. Tikuma Shobou, 筑摩書房. (1637). *La geometrie.*

Descartes, R., Miyake, T., & Koike, T. (trans. 1993), 方法序説. Hakusuisha, 白水社. (1637). *Discours de la methode.*

Galilei, G., & Aoki, Y. (trans. 1959-1961), 天文対話 上・下. Iwanami Shoten, 岩波書店. (1632). *Dialogue concerning the two chief world systems.*

Galilei, G., Konno, T., & Hida, S. (trans. 1937-1948), 新科学対話 上・下. Iwanami Shoten, 岩波書店. (1638). *Dialogue concerning two new sciences.*

Galilei, G., Yamada, K., & Tani, Y. (trans. 1976), 星界の報告. Iwanami Shoten, 岩波書店. (1610). *Sidereus nuncius.*

Gleick, J. (2003). *Isaac Newton.* London: Fourth Estate. A division of HarperCollins Publishers.

Kepler, J., & Kishimoto, Y. (trans. 2009), 宇宙の調和—不朽のコスモロジー. Kosakusha, 工作舎. (1619). *The harmony of the world* [*Harmonices mundi libri*]. Lincii Auftriae.

Newton, I., & Nakano, S. (trans. 1977), プリンキピア—自然哲学の数学的原理. Kodansha, 講談社. (1726). *Philosophiae naturalis principia mathematica.* 3rd ed.

Newton, I., & Shimao, N. (trans. 1983), 光学. Iwanami Shoten, 岩波書店. (1721). *Optics.* 3rd ed.

Sootin, H., & Watanabe, M. (supervised the translation), Tamura, Y. (trans. 1977), ニュートンの生涯. Tokyo Tosho, 東京図書. (1955). *Isaac Newton.* London: Julian Messner, a division of Simon & Schuster, Inc.

Sugget, M., & Oohasi, K. (trans. 1992), ガリレオと近代科学の誕生. Tamagawa University publishing department, 玉川大学出版部. (1981). *Galileo and the birth of modern science.* Wayland Publishers Ltd.

White, M. (1998). *Isaac Newton: The last sorcerer.* London: Fourth Estate. A division of HarperCollins Publishers.

Yukawa, H., & Tamura, S. (1955–1962). *Accepted theory of physics* (Vols. I–III), 物理学通論, 上・中・下. Tokyo: Taimeidou, 大明堂.

Chapter 2
Michael Faraday

Michael Faraday discovered electromagnetic induction, which led to the invention of the generator, the motor, and the transformer, vastly and rapidly improving the quality of life. He researched plenty of electromagnetic phenomena, such as electromagnetic induction, polarization of dielectrics, magnetization of substances, and the relationship of light and magnetism. James Clerk Maxwell gave his electromagnetic theory on the basis of experimental research by Faraday. Electromagnetic theory and Newtonian mechanics constituted the greatest two theories of classical physics until the end of the nineteenth century.

2.1 Upbringing

Birth of Michael Faraday

Faraday's father, James Faraday, was a blacksmith working for an ironmonger James Boyd. In 1787, he and his wife, Margaret, relocated to Butts, near London, on Boyd's recommendation. On May 26 the same year, their first child, daughter Elizabeth, was born, and on October 8 the next year, they had a son, Robert. Three years later, on September 22, 1791, their second son, Michael Faraday, was born in Newington, Butts, which was then at the edge of London. Beyond lay the fields of Surrey. Butts is part of London nowadays. Faraday was named after his maternal grandfather, Michael Hastwell.

When Faraday was five years old, the family relocated to rooms over a coach house in Jacob's Mews near Welbeck Street in west London, where his father worked. His father, in his mid-forties then, suffered from such ill health that he could not work long hours and found it difficult to earn enough to feed his family. In 1801, the price of bread peaked because of shortages occasioned by the war with France (James, 2010, p. 10), and the family was badly off. So the family received public relief. However, at the beginning of the nineteenth century in Great Britain, public facilities such as workhouses treated poor people badly. Faraday's family probably faced insolence because they were poor. In fact, Faraday was authorized only a loaf of bread per week. His mother took in lodgers for a living.

His education was most ordinary, consisting of little more than the rudiments of reading, writing, and arithmetic at a common day-school. His hours out of school were passed at home and in the streets (James, 1991, p. xxvii).

This period saw massive progress on one hand and equally massive social unrest on the other. In Great Britain, the industrial revolution, which occurred largely in Manchester and Liverpool in Lancashire, was in progress. Under King George III, London was the largest, richest, and most powerful city in the world.

Simultaneously, in the Continent, the French Revolution, which began in 1789, was in progress. In 1793, the monarchy was abolished, and Louis XVI and Queen Marie Antoinette were executed. After the revolution, France was opposed to Great Britain. Napoleon Bonaparte, who became emperor, antagonized Great Britain. In 1837, the Victorian Age commenced in Great Britain.

Apprenticeship

To assist his poverty-stricken family, Faraday at the age of 13 started working. On September 22, 1804, his birthday, he was employed as a newspaper-cumerrand boy at the bookshop of bookseller Frenchman George Riebau at 2 Blandford Street, near his house. His work was to deliver the bound books, newspapers, and magazines. He also had to collect the rented magazines—subscribing to the magazine was expensive, so most people rented it. Faraday had busy days.

Later in his life, whenever he would meet a newspaper-cumerrand boy, he always spoke kindly to him. He told his niece that because he had experience carrying newspapers, he was kind to other newspaper boys (James, 2010, p. 20).

Riebau was a first-class bookbinding artisan. Generally, persons working under such artisans learned bookbinding as apprentices. To be an apprentice, a fee had to be paid. But Faraday could not pay the fee because earlier, his brother was apprenticed to a blacksmith, and his father could not afford to pay for Faraday's apprenticeship. However, Riebau loved Faraday as a son and allowed Faraday to become apprenticed to him when he turned 14 years of age. On October 7, 1805, the indenture for apprenticeship was made. The indenture described the terms that instead of paying a fee, Faraday should work faithfully and avoid taverns. After completing an apprenticeship of seven years, the apprentice became an artisan and got wages. Faraday was very skillful and learned excellent bookbinding.

Faraday took delight in reading various scientific book (including Jane Marcet's *Conversations on Chemistry* and the electrical treatise by James Tytler in the third edition of *Encyclopaedia Britannica*) (James, 1991, p. xxix) bound beautifully by him. Garnering scientific knowledge by reading books, he became interested in science. For Faraday, with only an elementary education, books were his only teacher. He performed chemical experiments to confirm the knowledge obtained from books, using his free time at home after work. He fabricated a device causing friction electricity, which is now held in the Royal Institution.

From February 19, 1810, to September 26, 1811, during apprenticeship, Faraday attended lectures by John Tatum, the leader of the City Philosophical Society in London. This society had been founded in 1808 to give artisans and apprentices, like Faraday, additional access to scientific knowledge. There were many such lectures and societies in London and provincial cities during the first quarter of the nineteenth century. Those who attended such societies believed that self-improvement would lead to better prospects for themselves both materially and morally—materially because it would continue to help with the ever-increasing pace of the industrialization of Britain and morally because as Faraday himself put it, philosophical men had superior moral feelings (James, 1991, p. xxix). Tatum delivered 12 lectures during this period. Among these, seven lectures were concerning electricity. Faraday's brother, who had become a blacksmith but had always understood Faraday's interest in science, paid 12 shillings as fee for the 12 lectures. On October 30, 1810, their father passed away and the brother took over the care of the family.

Lectures by Tatum gave advanced scientific knowledge. Experiments on Leiden's bottle and the electrochemical decomposition of salts were shown. These lectures provided Faraday wide scientific knowledge, equivalent to that gained in higher education.

When Faraday could not understand any content of a lecture, he would immediately consult *Encyclopaedia Britannica* for clarification. He was unusually eager to garner knowledge. He completed four notebooks with knowledge gained from the lectures. These notebooks also indicated how Faraday who could not receive regular schooling due to poverty yearned a scholarship.

While attending the lectures, he made many true friends. Among them was Richard Phillips, who formed lifelong friendship with Faraday and afterward became an excellent chemist.

2.2 Davy, Professor of Chemistry at the Royal Institution

Lecture by Davy

Among the customers at Riebau's bookshop, there was William Dance's son, who was a subscriber and life member of the Royal Institution. The encounter of Faraday with Dance was a turning point in his life.

Dance had the opportunity to study Faraday's lecture notebooks. The notebooks indicated that Faraday had understood the contents of all lectures completely. Dance was astonished at Faraday, who having received no higher education still had an understanding of such advanced scientific knowledge, and thought that Faraday had an extraordinary brain.

In 1812, Dance presented Faraday, then 21 years old, a ticket to attend a lecture by Humphry Davy, professor of chemistry at the Royal Institution. In December 1807, the Académie des Sciences awarded Davy the Volta Prize, worth 3,000 francs, instituted by Napoleon in an effort to antagonize Great Britain. Davy was the most outstanding scientist in Europe.

Davy delivered the lecture along with experiments, such as an experiment on electrochemical decomposition. In those days, acids were considered to contain oxygen. But Davy indicated that muriatic acid was a compound of hydrogen and chlorine and contained no oxygen. The experimental equipment that Davy used, Faraday had only seen in books and marveled at the use of such equipment by a professional.

Faraday found Davy's lecture, which was the final lecture by the professor in the Royal Institution, interesting and precisely summarized it in his notebook. On April 8, 1812, Davy was knighted by the Prince Regent (James, 2010, p. 31), and three days later, a very wealthy widow, Jane Apreece, became his wife. Because he no longer needed to work due to her wealth, at the age of 34, he retired from his professorship at the Royal Institution. But the Royal Institution's manager was anxious to retain their connection with the most famous English chemist of the

Fig. 2.1 Humphry Davy (1778–1829). (Portrait made in 1803.)

time and appointed Davy honorary professor of chemistry and experimental chief. Consequently, Davy had great influence in the Royal Institution even after marriage.

— ·⟫ ⟪· — ·⟫ ⟪· — ·⟫ ⟪· —

Explanation 2.1 The Royal Institution of Great Britain

In 1799, at 21 Albemarle Street, the Royal Institution was founded (Figs. E2.1 and E2.2). The purpose was "useful mechanical inventions and improvement, the application of science to the common purpose of life, and diffusing the knowledge by courses of philosophical lectures and experiments" (James, 2010, p. 27). Subscribers on foundation became life members.

Fig. E2.1 The Royal Institution.
(Painting made in 1838.)

Fig. E2.2 The Royal Institution.
(Photograph taken by Dr. K. Matsuda in September 2017.)

In those days, explosions were common at collieries, mainly because the fire-damp (a flammable gas) in the collieries caught fire from the miners' lamps. The safety lamp invented by Davy addressed this hazard and helped the lives of plenty of workers saved. Plenty of researchers, including Davy and Faraday, at the Royal Institution made scientific contributions. For example, James Dewar (Fig. A2.2) pioneered low-temperature science and succeeded in liquefying hydrogen gas, and he invented the Dewar bottle, which preserved the liquefied gas in extremely low temperatures. The father-and-son duo William Henry Bragg (1862–1942) and William Lawrence Bragg (1890–1971) researched in the Royal Institution. Thus, for about 200 years, the Royal Institution produced genius researchers contributing to development in science. Since 2007, it has not been used as a place for research. In 1973, Faraday Museum opened in the Royal Institution.

—⟫ ⟪— ⟫ ⟪ —⟫ ⟪—

Yearning to be a scientist

Since attending Davy's lecture, Faraday had yearned to be a scientific practitioner. For him, the Royal Institution was the most fascinating work place.

In October 1812, when he was 21 years old, his apprenticeship ended and he started his career with De La Roche. The new proprietor was whimsical and not a person who encouraged Faraday's enthusiasm for studying. Faraday often returned home late and found it difficult to find time for his experiments. As a result, his desire to get a position at a scientific workplace became stronger.

In 1812, he decided to send Joseph Banks, president of the Royal Society, a letter applying for a position at a scientific experimental room. In his letter, he asked to be "engaged in scientific occupation, even though of the lowest kind" (James, 1991, p. xxx). From his memory of delivering books from Riebau's bookshop, he found Banks' residence. He asked the gatekeeper to pass on the letter to Banks and said that he would return later for the reply later. As promised, he returned one week later but was handed back just his own envelope, on which "no reply" was scribbled. Afterward, he kept seeking jobs but was always rejected. The reason was that he had neither an academic career nor any qualification. Everyone inquired after what he knew and could do. Faraday realized that a person with neither an academic career nor any qualification could not get a post at a scientific workplace and lost confidence completely (Sootin *et al.*, 1976, p. 35).

The number of scientific practitioners in Great Britain was small then. Most practitioners were earning their living in other professions or possessed private wealth. With a social background like his, it was difficult for Faraday, without private wealth, to get a job.

Since he had relocated to De La Roche, he used to often visit Riebau's bookshop—Riebau was very kind and helpful to Faraday. One day, when he visited the bookshop, Dance happened to be there. Dance suggested that Faraday send a letter applying for a post in the Royal Institution directly to Davy, attaching his notebook summarizing the lecture by Davy. Faraday told him that he had sent a letter to Banks but had received no reply. However, Dance persuaded him to try again, to which Faraday agreed.

Davy, who received Faraday's letter and notebook, showed the letter to a director John Pepys and sought his advice, saying, "A youth called Faraday asks to be employed in the Royal Institution. What can I do?" Pepys is said to have answered, "Let him wash test tubes. If he is a valuable person, he will wash them obediently. If he refuses, he is not worth supporting." To this answer, Davy replied that he could not examine the youth in this manner and desired to judge him by a better method (Tyndall, 2002, p. 2). Davy at the time was honorary professor of chemistry in the Royal Institution and the head of laboratory.

On Christmas Eve in 1812, one week after Faraday posted the letter and the notebook, the reply from Davy was delivered to him. It was a letter acknowledging Faraday's serious attitude toward science. In the letter, Davy notified that he was leaving London and would meet him while returning in January 1813.

Interview with Davy

Early in January next year, Faraday was informed about the time and date of the meeting following Davy's promise. It was arranged that Faraday would meet with Davy in the Royal Institution. Faraday went to the Royal Institution with a feeling of expectation mingled with anxiety.

During the interview, Davy inquired why Faraday wished to be a scientific practitioner. Faraday explained his academic enthusiasm. Davy said that science is a harsh mistress and pays poorly (James, 1991, p. 497). However, Faraday answered that scientists got their greatest reward from seeking the truth by researching science and that science lets humans have superior moral feelings (Sootin *et al.*, 1976, p. 51).

Davy inquired whether Faraday's knowledge concerning science was obtained by self-study. Faraday, who had received almost no regular education, thought that he would be inquired about his academic career and got nervous, remembering his previous sad experience in seeking a job. However, Davy did not inquire about his academic career and said that the tidy notebook summarizing the lecture by Davy indicated Faraday's enthusiasm for science, good memory, and capacity (Sootin *et al.*, 1976, p. 54).

Davy's career in youth was similar to Faraday's career. During his apprenticeship with an apothecary, Davy read *Traite Elementaire de Chemie* by Antoine-Laurent de Lavoisier and took a strong interest in chemistry. The hardships he faced in his boyhood were similar to those faced by Faraday, leading Davy to deal very kindly with Faraday.

At the end of the interview, Davy advised Faraday to continue as a bookbinding artisan as it was a reliable profession and notified him that there was at that time no vacant post in the Royal Institution. However, at parting, Davy said that when a post became available, he would remember Faraday (Sootin et al., 1976, p. 55). Davy's last words gave hope to Faraday. In fact, these words would change Faraday's life.

Three-day work at the Royal Institution

Two days after the interview, Faraday received a letter from Davy, who had just sustained injuries following an explosion caused by combining nitrogen and chlorine. Glass had penetrated Davy's eyes, and his vision was impaired. Because the deadline for submitting his paper was close, Davy informed Faraday by the letter his desire that Faraday go to the Royal Institution to make a fair copy within three days of the draft of experimental notes by Davy.

Faraday could easily understand all the academic terms described in the experimental notes and accomplished the work in three days. Davy was satisfied with Faraday's work because Faraday had the ability to understand scientific work and his handwriting was beautiful. For Davy, Faraday became a favorite.

This three-day work was an opportunity for Faraday. Had he not understood the academic terms and been able to make a fair copy of the draft, then he would have lost an excellent chance to prove his scientific acumen. Fortunately, Faraday made excellent use of the opportunity.

Later, one night in February, when Faraday was undressing for bed, a splendid coach pulled up at his house, 8 Weimous Street, where Faraday's family had lived since 1809. Davy's footman delivered a note requesting Faraday to call the following morning (James, 2010, p. 34).

John Newman, instrument maker to the Royal Institution, had accused Davy's assistant William Payne of assaulting him when he was delivering experimental apparatus to the Royal Institution and Payne was immediately dismissed. Faraday was offered the post left vacant by Payne. Davy, after all, had kept the promise he had made to Faraday at the interview earlier.

2.3 Opening the Doorway to Research

Journey to the Continent

On March 1, 1813, Faraday was appointed as an experimental assistant in the Royal Institution. The previous year, Faraday had served as Davy's amanuensis for a while but Davy had had no permanent position to offer to Faraday then.

On October 13, seven months after he had joined the Royal Institution, Davy departed with Faraday for a journey across the Continent. For Faraday, this journey was an opportunity to get extensive knowledge. They demonstrated chemical experiments and took interviews in various countries. They began their journey in France, and Davy showed the properties of iodine to chemists in Paris. In his correspondence, Faraday described that iodine is a heavy black element like lead. On being heated, it melts and becomes a beautiful purple gas. On being cooled, it becomes crystal. Iodine forms compounds with all metals except platinum and gold (James, 1991, p. 74). Davy submitted his paper summarizing the results of his experiments on iodine to the Royal Society. They interviewed André-Marie Ampère (Fig. 2.2). Ampere discovered Ampere's law concerning electric current seven years after the interview.

After visiting Lyon and Nice from Paris, they went across the 6,000-foot Alps, at the border France and Italy, to Italy, where they visited Turin, Genoa, and Florence.

Fig. 2.2 André-Marie Ampère (1775–1836).
(Portrait made in 1825.)

Fig. 2.3 Ponte Vecchio, Florence,
painted by Antonietta Brandeis (1848–1926).

In Florence, Davy conducted his experiment on combustion involving a diamond placed at focus of the lens. The diamond was placed in the middle of a glass globe, supported in a cradle of platinum. To start the combustion, the glass globe was pierced full of holes. The diamond placed at the focus of the lens was heated, and it burned with a beautiful, vivid scarlet light. Afterward, the glass globe was found to contain nothing but a mixture of carbonic and oxygen gases. This proved that diamond is pure crystalized carbon.

In his correspondence, Faraday said that Florence was beautiful and preserved enormous amounts of instructive things and that the city was like a fine museum of natural history. He also said that the telescope with which Galileo had discovered the satellites of Jupiter and the first lens Galileo had polished were in a museum in Florence (James, 1991, p. 75).

Fig. 2.4 Jean-Baptiste Andre Dumas (1800–1884). (Portrait made in 1840–1850.)

From Florence, Davy and Faraday went to Rome, Naples, and Milan. In Milan, they interviewed Alessandro Volta (Fig. A2.9), who had invented the electric battery. From Italy, they went to Switzerland. In Geneva, Faraday made friends with Jean-Baptiste Andre Dumas (Fig. 2.4), 15 years younger than him. Afterward, Dumas made contributions to organic chemistry and became a professor of chemistry at Ecole Polytechnique.

Davy and Faraday went from Geneva to Lausanne, Berne, Zurich, and Munich and across the Tyrol Alps, at the Austria-Italy border, back to Italy again and from there to Padua and Venice.

Napoleon's escape from Elba

Through Bologna and Florence, Davy and Faraday arrived in Naples. Their schedule was to go on to Constantinople, Turkey. But hearing of Napoleon's escape from Elba, they decided to end their journey, afraid of an unstable political situation. Again crossing the Tyrol Alps, they went to Germany, the Netherlands, and Belgium and sailed across to Great Britain. Finishing a long journey of one year and six months, on April 23, 1815, they arrived in London. On June 18, 1815, the Battle of Waterloo took place between the French forces commanded by Napoleon and the allied forces of Britain, Holland, and Prussia in Belgium.

Davy lamp

In the summer of the year Davy and Faraday came back to London, Davy was requested to create a miner's safety lamp that could light in a colliery without exploding because of the flame of the lamp coming in contact with firedamp such as methane gas, and Faraday assisted him in his study for two months, from October middle (Explanation 2.2). From 1818 to 1822, Faraday did joint research with James Stodart, a British surgical instrument maker, to improve the quality of steel. The aim was to produce a steel alloy with excellent hardness by adding a little platinum and nickel. The Royal Institution had excellent facilities. This is the reason an outside company requested the joint research to be conducted in the Royal Institution. Undertaking research on alloys, Faraday built his reputation as a chemist.

When conducting joint research with an outside company, Faraday would get extra income. As more and more joint research studies were conducted, Faraday's incidental income exceeded his original income. Had he continued like this, he could have accumulated a lot of wealth. However, he soon realized then in any sort of joint research, the person or enterprise offering the funds was the leader of the research and that he could not do research according to his own thoughts and ideas. So Faraday gradually started rejecting contracts offering funds for joint research and devoted himself to primary research at the Royal

Institution. Though he might have been badly off financially, he elected to have autonomy rather than money.

—◊〉〈◊— ◊〉 〈◊• —◊〉〈◊•—

Explanation 2.2 Davy lamp

While conducting research on miners' safety lamp, Davy verified that when the inner radius of the metal tube was less than 0.1428 inches and the depth was proportionally more in comparison to the inner radius, the mixture of air and flammable gas did not explode and even if an explosion occurred, it could not go through the tube. Heat loss from the surface of the metal with thermal conduction contributed to this phenomenon. On the basis of the above results, he fabricated the safety lamp (Fig. E2.3) as follows: To allow the flame of the lamp to illuminate the area, the side plane was made of four glasses, but the upper and bottom surfaces were made of metal. To allow the air needed to support the flame to enter, the bottom metal surface was pierced with small holes 0.125 inches in diameter and 1.5 inches deep. The upper metal surface was pierced with small holes to vent, playing the role of a chimney. Thus, the inside and outside of the lamp were linked via small, deep holes. That is, the flame on the inside and the flammable gas on the outside came in contact through only the small holes. Davy published a paper on the research results in 1816 (Davy, 1816).

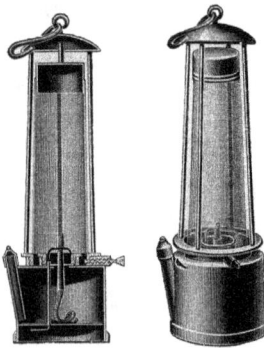

Fig. E2.3 The Davy lamp.

—◊〉〈◊•— ◊〉 〈◊• —◊〉〈◊•—

2.4 Oersted's Discovery

Phenomenon of electromagnetism

In 1820, Hans Christian Oersted (Fig. 2.6), in Denmark, discovered the phenomenon of electromagnetism. As shown in Fig. 2.5, when an electric current i passed through a linear conductor,

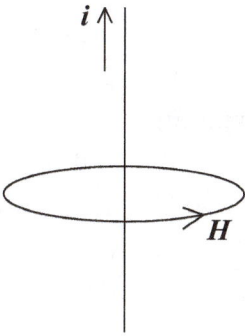

Fig. 2.5 Relation between current (*i*) and magnetism (**H**).

Fig. 2.6 Hans Christian Oersted (1777–1851). (Portrait made in 1832.)

magnetism **H** occurred in the same direction as the rotation of the screw, proceeding in the direction of the electric current. In other words, the magnetic needle facing south-north changed its direction to a direction perpendicular to the wire.

Oersted and the inverse problem

Oersted's experiment showed a connection between electricity and magnetism. He wrote his paper on this in Latin, but due to its importance, it was translated into all European languages. After Banks, William Hyde Wollaston (Fig. 2.7) was temporarily made president of the Royal Society, until Davy's appointment. When Wollaston heard of Oersted's experiment,

Fig. 2.7 William Hyde Wollaston (1766–1828). (Portrait made in 1820–1824.)

he decided to study the inverse problem of it. He thought that if a magnetic needle moved because of a wire through which electricity was passing, then the wire should also move because of the magnetic needle, in accordance with the action-reaction law. He tried to realize his idea in the Royal Institution in the presence of Davy. Faraday, seeing the experiment being carried out, became interested in the problem. Phillips, who had got acquainted with Faraday at the lecture by Tatum mentioned above, recommended that Faraday write a review of the paper by Oersted in *Annals of Philosophy*. So Faraday carried out many experiments on the problem.

Faraday grappled with the problem using a different method than Wollaston. As shown in Fig. 2.8, using mercury (Hg) as the liquid conductor, he set one end of wire B against hinge A, making a movable wire B, and dipped the other end of wire B into the mercury. He set a magnet M at the center of the container with the mercury. When he brought wires B and C in contact with the electric battery V, wire B began rotating around the magnet M. He called this phenomenon "electromagnetic rotation." In the motor mentioned in Section 2.7, the coil rotated when an electric current passed through a coil in a magnetic field. In Faraday's electromagnetic rotation, the

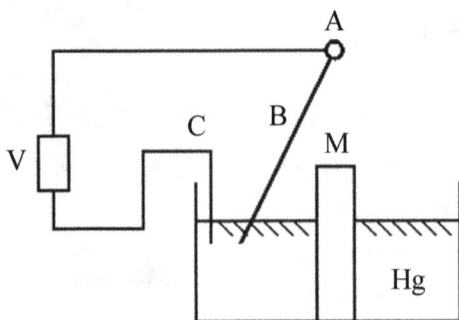

Fig. 2.8 Faraday's method to address the inverse problem of Oersted. A: hinge; B: movable wire; V: electric battery; M: magnet.

wire rotated when an electric current passed through a wire in a magnetic field, and therefore Faraday's electromagnetic rotation is regarded as a prototype of a motor.

However, people questioned the originality of Faraday's work. It was rumored that Faraday had stolen part of Wollaston's research (James, 2010, p. 39). Faraday published his paper without quoting Wollaston's experiment. Therefore, it seemed to many people that Faraday had imitated Wollaston's experiment without giving any recognition to Wollaston's work. In fact, the reason for Faraday's action was that Faraday could not connect with Wollaston and he did not desire to quote Wollaston's research without his permission (Bowers, 1978, p. 44). Though Wollaston declined to take any action against Faraday, people continued to blame Faraday.

The controversy reached its zenith when Faraday was nominated as Fellow of the Royal Society. It was reported that Davy thought that Wollaston was the first to propose the inverse problem of Oersted and had reservations about Faraday being the one to discover electromagnetic rotation (James, 2010, p. 39).

Marriage

On February 20, 1791, Faraday's father made his confession of faith in a Sandemanian church, a non-Anglican church. On July 15, 1821, Faraday made his confession of faith in a Sandemanian church, like his father. On June 12, one month before the confession, Faraday married Sarah Barnard (Fig. 2.9),

Fig. 2.9 Faraday and his wife.

who was Sandemanian and the daughter of a silversmith. She was nine years younger than Faraday and a modest woman. They lived in the attic of the Royal Institution, where Davy had lived in his youth, for the next 37 years.

Faraday was kind to her. She took care of him. Despite plain living, they lived a life filled with affection.

2.5 Liquefaction of Chlorine Gas

In 1823, Faraday liquefied a gas (chlorine) for the first time. As shown in Fig. 2.10, he used a Bunsen burner to heat a chloric compound at one end of a sealed inverse V-type glass tube. The sealed glass tube filled up with a yellow chlorine gas, produced by the decomposition of the compound, creating a high pressure within the tube. Under the pressure, the gas liquefied and collected at the other end of the tube, which was at a low temperature. The experiment indicated that the gas was the vapor of the liquid with a low boiling point. This is how Faraday became aware that high pressure and low temperature played an important role in the liquefaction of a gas (the historical significance of the liquefaction of chlorine gas by Faraday is described in Appendix 2.1). Questions were raised regarding the originality of Faraday's success in liquefying chlorine. Davy believed that it was he who had given the suggestion to Faraday to carry out the experiment and so he should have received the honor of discovery of liquefaction of a gas. The controversy

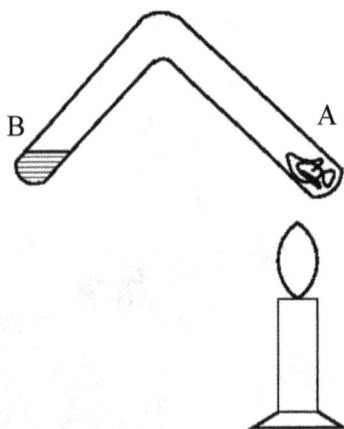

Fig. 2.10 Liquefaction of chlorine gas. A: chloric compound; B: liquefied chlorine.

with Davy reached the same level, creating as much bad feelings, as the one on the inverse problem of Oersted, when Faraday was nominated as Fellow of the Royal Society.

—❯❯❮❮•— •❯❯ ❮❮• —❯❯❮❮•—

Explanation 2.3 Liquefaction of gas

The Fig. E2.4 phase diagram expresses the state of matter at pressure P and absolute temperature T. The vertical axis represents pressure P, and the horizontal axis represents the absolute temperature T. The unit of absolute temperature is Kelvin (K) and is expressed in centigrade (C), as follows:

Absolute temperature (K) = 273.15 + centigrade (C).

Because absolute temperature is greater than 0 K, only temperatures above 0 K can be used using absolute temperature.

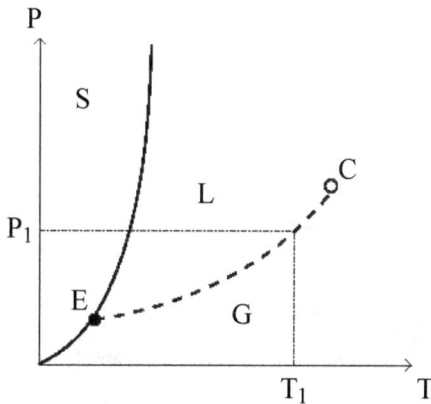

Fig. E2.4 Phase diagram. S: solid; L: liquid; G: gas; C: critical point; E: triple point; P: pressure; T: absolute temperature.

The solid line in Fig. E2.4 represents the boundary between the states of matter. The solid line above the triple point E represents the boundary between solid and liquid. Here solid equilibrates with liquid. This shows the melting point of the solid. The solid line below the triple point E represents the boundary between solid and gas, and here solid equilibrates with gas—this is where solid becomes gas, or sublimation occurs.

The broken line in Fig. E2.4 represents the boundary between liquid and gas, and here liquid equilibrates with gas. This shows the boiling point of the liquid. The upper end, C, of the broken line represents the critical point, and above this point, the surface that distinguishes liquid from gas is extinguished. When the pressure is increased at temperature T_1 below the critical point, as shown in Fig. E2.4, the gas liquefies. At a temperature above the critical point, increasing pressure does not induce liquefaction of the gas. The broken line is called the vapor-pressure curve.

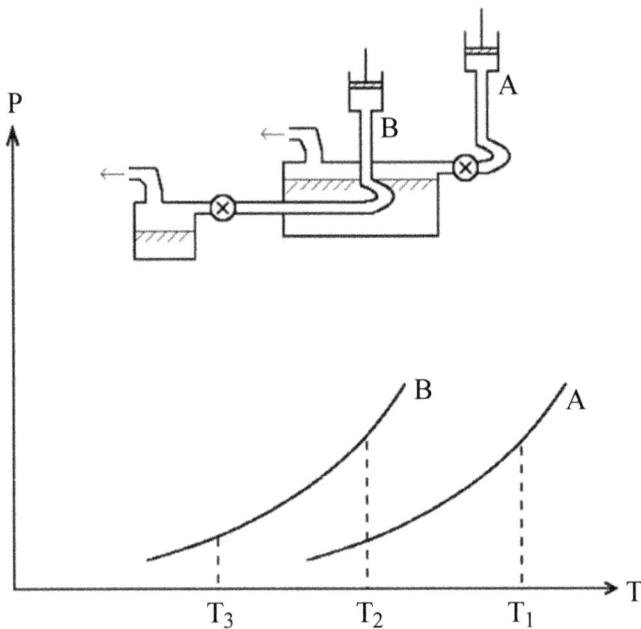

Fig. E2.5 Multistage cooling. P: pressure; T: absolute temperature.

The multistage cooling process is described in Fig. E2.5. Matter A liquefies at T_1 below the critical point under increasing pressure. The gas in the vessel of liquefied matter A is evacuated by a vacuum pump. As a result, the pressure decreases and the temperature of matter A falls according to the vapor-pressure curve. As the temperature decreases further, matter B liquefies at T_2 below the critical point by compression. By decreasing the pressure by evacuating the gas from the vessel of liquefied matter B, matter B can be cooled till its temperature reaches T_3. Cooling like this is called multistage cooling. Heike Kamerlingh Onnes (Fig. A2.1) liquefied air with methyl chloride, ethylene, oxygen, and air using multistage cooling.

In 1908, while preparing liquefied air and liquefied hydrogen, Onnes succeeded in liquefying helium gas using the Joule–Thomson effect. First, he liquefied hydrogen gas, cooled in advance by liquefied air. Next, using the Joule–Thomson effect, he liquefied helium gas cooled in advance by liquefied hydrogen. The Joule–Thomson effect is the phenomenon when if gas is streamed through a tube containing padding with plenty of holes and is expanded, the temperature of the gas changes. Near liquefaction, the cooling is remarkable in the Joule–Thomson effect.

—◦}}⟨⟨◦— ◦}} ⟨⟨◦ —◦}}⟨⟨◦—

Appendix 2.1 Historical significance of liquefying chlorine gas

Faraday's knowledge about high pressure and low temperature to liquefy gas, was used in a difficult process but important for realizing extreme low temperatures half a century after Faraday's liquefaction. For example, at Craco in Poland, Zygmunt Wroblewski and Karol Olszewski succeeded in liquefying oxygen in 1883 by introducing oxygen gas under high pressure into a tube that was cooled by being soaked in liquefied boiling ethylene and then evacuated. In 1908, when Onnes, in Leiden University in Netherlands, succeeded in liquefying helium gas, which was most difficult to liquefy, he utilized the multistage cooling method of Wroblewski and Olszewski. Thus, Faraday's knowledge of high pressure and low temperature played an important role in the liquefaction of gases.

Onnes was not satisfied with his success in liquefying helium gas, and under the condition of the extremely low temperature realized by liquefied helium, he tried to research property of matter. On experimenting with liquefied helium, he discovered a phenomenon called "superconductivity,"

Fig. A2.1 Heike Kamerlingh Onnes (1853–1926).

where the resistance of mercury became zero under transient temperature. Superconductivity was an important discovery in the twentieth century.

Because resistance of matter in the superconductive state is zero, the thermal loss is zero on shedding current and so a great amount of current can be shed. Then an electromagnet with a coil of superconductive matter can produce a great magnetic field, and at present, an electromagnet using superconductivity is utilized in the linear super express, Japan, and in magnetic resonance imaging machines (Shioyama, 2002).

Fig. A2.2 James Dewar (1842–1923).

In the history of technology of liquefying gases, which was an important method realizing extreme low temperatures, Faraday's liquefaction of chlorine was the pioneer. Dewar, who succeeded in the liquefaction of hydrogen gas, always respected Faraday, who had accomplished the liquefaction of chlorine gas in the same institution (Mendelssohn & Ooshima, 1971).

The cryogenics utilizing cryogenic technology play an important role in modern society. The history of cryogenics goes back to Faraday's works.

—◊◊◊◊•— ◊◊ ◊◊• —◊◊◊◊•—

2.6 Election for Fellow of the Royal Society

Nomination by Phillips

In 1823, Faraday was nominated for Fellow of the Royal Society. This was very honorable for any scientist. In 1822, chemist Phillips, who was a friend of Faraday since they had attended the lecture by Tatum, was already Fellow. Phillips nominated Faraday.

Though there was no rule, in accordance with the precedence, the proposer usually consulted with the president before making any nomination. Davy was the president of the Royal Society at the time. But without consulting him, Phillips nominated Faraday. Davy got angry and asked Faraday to refuse the nomination. Faraday declined, saying that only the proposer could refuse the nomination (James, 2010, p. 40).

Banks, the preceding president, had elected many persons who were not scientists by any means, utilizing his position. Since Davy was made president, he had endeavored to reform this. There was discontent among some parties in the Royal Society regarding the strict reforms concerning the election of Fellow. But Davy was anxious for a good opinion in the Royal Society. The official reason of Davy's opposition to electing Faraday was as follows: "Davy did not desire for him to be thought to agree nominating Faraday favoring his own disciple" (James, 2010, p. 40). On January 8, 1824, Faraday was officially elected as Fellow of the Royal Society. There was only one opposition to Faraday's election, that of Davy's. Faraday was worried about Davy's intention. In February 1825, Faraday was promoted to the head of the laboratory in the Royal Institution. This promotion was proposed by Davy, which eased Faraday's worry about Davy's intention. When Davy was away from the Royal Institution during one of his journeys to the Continent and the islands of Great Britain, he made Faraday agent in his place.

In 1825, Faraday discovered a new substance. He discovered that the substance consisted of two constituents, found the ratio between the two, and called the substance "bi-carburet of hydrogen compound." Later, Eilhard Mitscherlich called the substance "benzene" (Appendix 2.2).

—❯❯❮❮—❯❯ ❮❮ —❯❯❮❮—

Appendix 2.2 Benzene

In 1865, Friedrich August Kekule von Stradonitz proposed the structure of benzene (Fig. A2.3), consisting of carbon and hydrogen (Kekule, 1865, pp. 98–110). According to Stradonitz, there is a hexagonal ring (Fig. A2.3), carbon atoms are placed at the corners of the hexagonal ring, each carbon atom is connected to the neighboring carbon atoms with a single bond on one side and a double bond on the other, and one hydrogen atom is connected to each carbon atom. For example, styrene, which is a raw material for synthetic resin, and synthetic rubber have benzene rings. This benzene ring would later play an important role in organic chemistry.

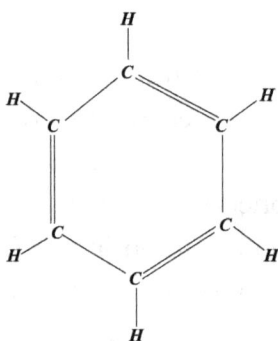

Fig. A2.3 Structure of benzene.

—❯❯❮❮—❯❯ ❮❮ —❯❯❮❮—

Friday evening discourses and Christmas lecture

From 1825, Faraday started delivering Friday evening discourses. Only the Royal Institution members and their guests could attend the lecture. Next year, he started to deliver the Christmas lectures. In 1861, his final Christmas lecture was published as *The Chemical History of a Candle* (Faraday & Takeuchi, 2010). This book was the most popular among the books published then. In 1854, when Faraday delivered a lecture on education, Queen Victoria's husband, Prince Albert, attended. Prince Albert, who was interested in the application of science, started attending Faraday's lectures frequently, and in 1855, he attended the Christmas lecture with his two sons.

Faraday delivered 127 Friday evening discourses and 19 Christmas lectures (Fig. 2.11). As writer George Eliot commented, Faraday's lectures explained state-of-the-art science in an entertaining manner, like an opera. Citizens listening to his lecture were attracted to his personality. The Friday evening discourses and Christmas lectures still continue in the Royal Institution.

Fig. 2.11 Faraday delivering a Christmas lecture. (Painting made in 1855 by Alexander Blaikley.)

2.7 Discovery of Electromagnetic Induction

Transforming electricity into magnetism

In 1825, William Sturgeon made an electromagnet by winding a coil around a piece of soft iron, giving mankind a method of transforming electricity into magnetism.

In the 1820s Faraday began thinking about how reverse transformation can be realized and wrote down his idea of "transforming magnetism to electricity" in his notebook, grappling with this problem because until then, no solution had been found to this problem yet.

Though in 1827 he was invited as professor of chemistry in the University of London, he declined and decided to research the great problem in the Royal Institution. In May 1829, Davy passed away in Geneva. The friendship between Davy and Faraday had ended when Faraday was elected as Fellow of the Royal Society, because of the controversies regarding the originality of research on the inverse problem of Oersted and the experiment involving liquefaction of chlorine gas.

Electromagnetic induction

For years, Faraday carried a bar magnet and a coil of wire in his pocket, constantly mulling over the problem of transforming magnetism into electricity. On October 17, 1831, he discovered electromagnetic induction—that temporal change of magnetism induces an electric current. Faraday submitted his first paper on electromagnetic induction on November 24, 1831 (published in 1832, Figs. 2.12 and 2.13) providing mankind a method of transforming magnetism to electricity that was inverse of the electromagnet created by Sturgeon. Electromagnetic induction

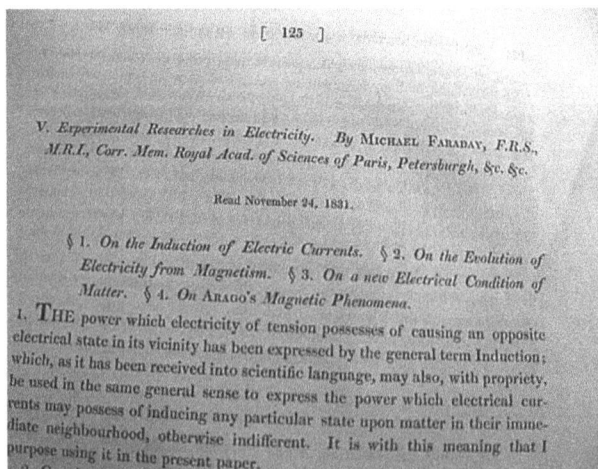

[125]

V. *Experimental Researches in Electricity.* By Michael Faraday, F.R.S., M.R.I., Corr. Mem. Royal Acad. of Sciences of Paris, Petersburgh, &c. &c.

Read November 24, 1831.

§ 1. *On the Induction of Electric Currents.* § 2. *On the Evolution of Electricity from Magnetism.* § 3. *On a new Electrical Condition of Matter.* § 4. *On Arago's Magnetic Phenomena.*

1. THE power which electricity of tension possesses of causing an opposite electrical state in its vicinity has been expressed by the general term Induction; which, as it has been received into scientific language, may also, with propriety, be used in the same general sense to express the power which electrical currents may possess of inducing any particular state upon matter in their immediate neighbourhood, otherwise indifferent. It is with this meaning that I purpose using it in the present paper.

Fig. 2.12 Faraday's paper on electromagnetic induction (Faraday, 1832, pp. 125–162).

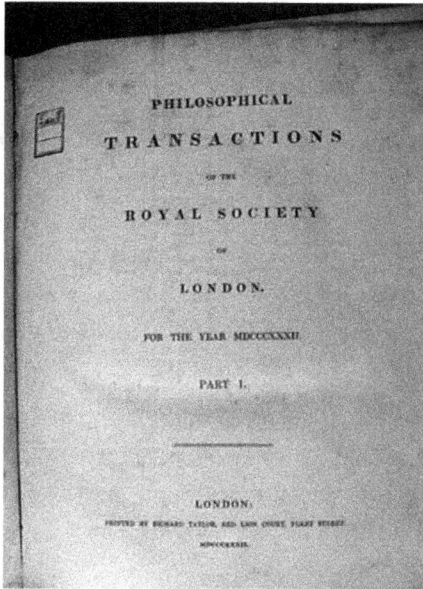

Fig. 2.13 Academic journal carrying a paper on electromagnetic induction (Faraday, 1832).

had induced the inventions of the motor, the generator, and the transformer, the precursors to modern conveniences. In a generator, a temporal change of magnetism at the coil rotating in a magnetic field by the power of water or vapor induces an electric current. On the other hand, in a motor, the rotation of a live coil in a magnetic field induces power. A motor is utilized in widespread fields, such as household electric appliances (like washing machines), communication instruments (such as computers), vehicles, and industrial instruments (such as lifts).

Faraday researched on the universal law of electromagnetic induction and concluded that it is necessary for a wire to cross a magnetic line force for a temporal change of magnetism to occur in order to induce inductive voltage. Inductive voltage is proportional to the number of crossing magnetic lines of force, whether the wire crosses the magnetic line of force perpendicularly or obliquely.

—◦⟩⟩ ⟨⟨◦— ◦⟩⟩ ⟨⟨◦ —◦⟩⟩ ⟨⟨◦—

Explanation 2.4 Experiment with an induction ring

Faraday conducted a number of experiments related to electricity and magnetism. In one of his experiments, as shown in Fig. E2.6, Faraday wound one coil on one side of a soft iron ring, the battery was connected to the coil, and the current could be switched on or off. At the other side of the ring, he wound another coil and made a closed circuit with a wire connected to the second coil. Near the wire connected to the second coil, a magnetic needle was set. If a current was induced, then the magnetic needle should have swung according to the phenomenon discovered by Oersted. Faraday thought that magnetism was induced in the soft iron ring when current was passed through the first coil and when the magnetism traveled to the part of the soft iron within the second coil, it might induce a current in the second coil.

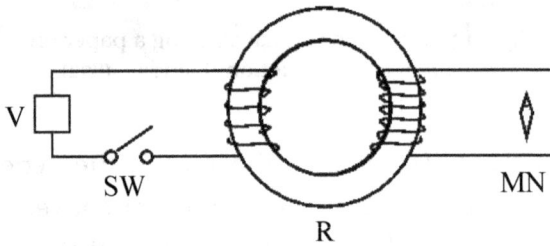

Fig. E2.6 Electromagnetic induction. R: induction ring; MN: magnetic needle; SW: switch; V: battery.

He noticed then when he opened the switch, the magnetic needle set near the wire connected to the second coil swung slightly. Again, when he closed the switch, the magnetic needle swung slightly. However, if a scientific fact is not reproduced under the same conditions, then the fact is not recognized as truth. That is, reproducibility must be verified. Faraday verified the reproducibility.

He became aware of the significant fact that the magnetic needle swung whenever the circuit was opened or closed (James, 2010, p. 57). He concluded that only when the stage of current in the first coil changed a current was induced in the second coil.

The change of current in the first coil introduced a change of magnetism in the soft iron ring. The change in magnetism induced a current in the second coil. He estimated that a magnetic change by any other method should also induce a current.

On October 17, 1831, Faraday conducted another experiment. As shown in Fig. E2.7, he wound a linear coil and near the wire connecting to the coil for making a closed circuit, he set a magnetic needle that played the role of detecting current. Above the central axis of the coil, he moved a magnet toward and away from the coil. The magnetic needle swung as he

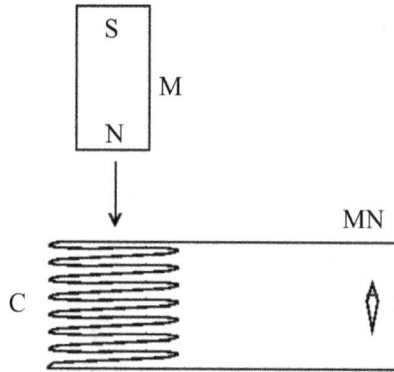

Fig. E2.7 A magnet's relative movement to a coil. C: coil; M: magnet; MN: magnetic needle.

had predicted. He found that intense movement of the magnet toward and away from the coil induced a large swing in the magnetic needle. That is, it was found that to induce electricity, it was necessary for the magnet to move and a temporal change of magnetism induced a current.

—◦‹‹‹◦— ◦‹‹ ‹‹◦ —◦‹‹‹◦—

Appendix 2.3 Magnetic and electric lines of force

Mathematical physicists in the Continent did not agree with Faraday's concept of a field of "lines of force" such as a magnetic line of force or an electric line of force. Instead, they agreed with André-Marie Ampère, whose electrodynamics took the form of remote action in mathematical formulation. Only Maxwell (Explanation 2.5) agreed with Faraday's concept. Immediately after graduating from the University of Cambridge, he undertook mathematicalization of Faraday's concept of lines of force. In 1855, he submitted his paper "On Faraday's Lines of Forces" to the Cambridge Philosophical Society, and it was published the next year. Maxwell recognized Faraday's research as "the nucleus of everything electric since 1830" (James, 2010, p. 89).

Furthermore, Einstein said, "the electric field theory of Faraday and Maxwell represents probably the most profound transformation of the foundations of physics since Newton's time" (James, 2010, p. 90).

—◦⟩⟩ ⟨⟨◦— ◦⟩⟩ ⟨⟨◦ —◦⟩⟩ ⟨⟨◦—

Explanation 2.5 Theorization of electromagnetic phenomena by Maxwell

In 1864, Maxwell (Fig. E2.9) derived electromagnetic equations expressed as four partial differential equations unifying all electromagnetic phenomena. One of them corresponds to electromagnetic induction, discovered by Faraday. The first term of the equation expresses spatial change of an electric field **E**, and the second term expresses a temporal change in the magnetic flux density **B**. $\mathbf{B} = \mu\mathbf{H}$, where $\mathbf{H}$ = magnetic field and μ = magnetic permeability, is also called magnetic induction. Assume that a coil of one turn is formed with a wire and a magnet is brought near it rapidly. When each term is integrated for the area of the interior coil, the first term becomes a line integral along the coil of the electric field and expresses inductive voltage. The second term expresses the temporal change of sum of the magnetic flux density penetrating the coil. That is, the above equation is the formulation of the electromagnetic induction discovered by Faraday. The other three equations are as follows:

- The occurrence of a magnetic field due to a temporal change in the electrical displacement $\mathbf{D} = \varepsilon\mathbf{E}$, where ε = dielectric constant and current.

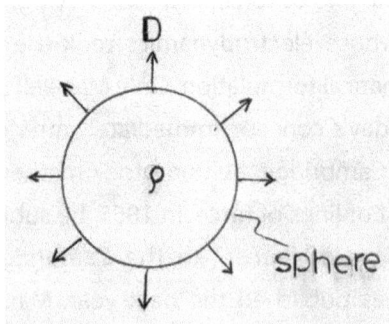

Fig. E2.8 Relation between **D** and ρ. The sum of **D** going out from a sphere = the sum of charges in the sphere.

Fig. E2.9 James Clerk Maxwell (1831–1879).

- The relation between electrical displacement and charge (when we take the volume integration of Maxwell's equation div $\mathbf{D} = \rho$ [ρ is the density of the electric charge], by Gauss theorem, the left-hand side becomes the surface integration of the component of $\mathbf{D}$ perpendicular to the surface and the right-hand side becomes the sum of electric charges in a volume. Therefore, the sum of electrical displacements coming out from a sphere is equal to the sum of the electric charges in the sphere) as shown in Fig. E2.8.
- The equation of continuity of magnetic flux density $\mathbf{B}$ (or the sum of magnetic flux densities going out from a sphere is equal to the sum of magnetic flux densities coming into a sphere because the volume integration of Maxwell's equation div $\mathbf{B} = 0$ becomes the surface integration of the component of $\mathbf{B}$ perpendicular to the surface. Consequently, the sum of components of $\mathbf{B}$ perpendicular to the surface of a sphere is 0). Thus, Maxwell systematized all electromagnetic phenomena (Yukawa & Tamura, 1955–1962). The curves expressing directions of electrical displacement and magnetic flux density are called electric line of force and magnetic line of force, respectively.

— ·》》 《《· — ·》》 《《· — ·》》 《《· —

Invention of a magnetic generator

Faraday invented a magnetic generator (Fig. 2.14) that received continuous direct current in a radial direction connecting

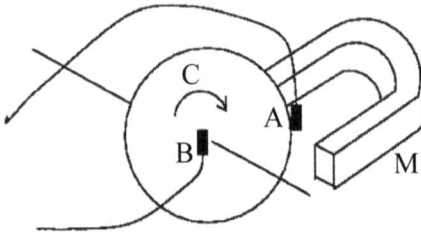

Fig. 2.14 The magnetic generator by Faraday. M: magnet; A, B: terminals sliding contacting; C: rotating copper board.

A and B on a rotating copper board. This was the prototype of a dynamo (direct current generator with a commutator). In an alternating current generator, the direction of the current is inverted every half rotation of the coil in a magnetic field. To always get direct current in a constant direction, in a dynamo, two fixed terminals are in sliding contact with a rotating commutator and are connecting with an outer circuit, and by changing the connection to the coil every half rotation, direct current moves in a constant direction to the outer circuit. The commutator consists of two metal boards, each of which is placed on a rotating shaft for two boards to go around the rotating shaft.

Electromagnetic induction, discovered by Faraday, is a phenomenon where if the magnetism changes at a place with an electronic circuit, such as a coil, that is, if the state of the magnetic line of force changes, power is induced. We call the power "inductive voltage" (Fig. A2.5) and call the current "inductive current." As explained in Appendix 2.4, the law of electromagnetic induction is derived on the basis of the behavior of electrons in a piece of metal moving in a magnetic field.

—❯❯❮❮•— ❯❯ ❮❮ —❯❯❮❮•—

Appendix 2.4 Modern interpretation of electromagnetic induction

An electron with charge −e moving at a velocity **v** in a magnetic flux density **B** receives a force **F**, called Lorentz force, as shown in Fig. A2.4. The direction of the Lorentz force is perpendicular to the plane containing **v** and **B**, and the magnitude of the Lorentz force is $evB\sin\theta$, with the sinusoidal function of angle θ between **v** and **B**. Here v is the magnitude of the velocity and B is the magnitude of the magnetic flux density.

In metal, there are free electrons, negatively charged, that can move freely and there are ions, positively charged, that cannot move freely (an ion is an atom in a crystal from which a free electron has been removed). When the wire moves at a velocity **v** in a field of magnetic flux density **B**, as shown in Fig. A2.5, the electrons in the wire are subjected to a Lorentz force. Because the free electrons move to the end Q of the wire, at point P of the wire, positive charges accumulate and at point Q of the wire, negative charges accumulate, and an electric field **E** (the force influencing the unit charge, i.e., the negative spatial gradient of potential) for the electrons in the direction Q to P is generated.

Because the electric field prevents the electrons from streaming nonstop to the Q end because of the Lorentz force, there is a state of equilibrium between the Lorentz force and the electric field. That is, the force $e\mathbf{E}$ due to the electric field for electrons and the Lorentz force $e\mathbf{B}\sin\theta$ remain in balance. Because the magnitude of the electric field is the spatial gradient of voltage, the magnitude of the electric field is given by the voltage between P and Q divided by the length d between P and Q. Because voltage is the

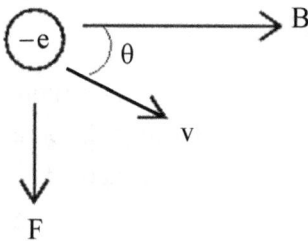

Fig. A2.4 Lorentz force.

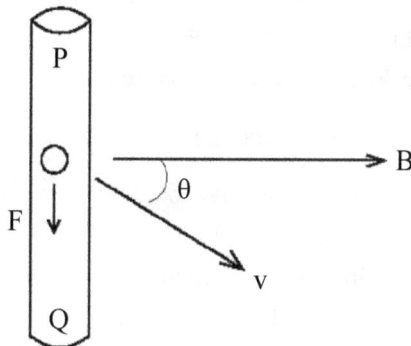

Fig. A2.5 Inductive voltage.

product of length d and the magnitude of the electric field, voltage is given by $d\mathbf{v}\mathbf{B}\sin\theta$.

If a wire is connected to a resistor R or a wire, then a current i flows (Fig. A2.6). Assume that the wire PQ in Fig. A2.7, perpendicular to the parallel sides of the U-shape wire, moves at a velocity $\mathbf{v}$ and $\mathbf{v}$ is perpendicular to $\mathbf{B}$ and the direction of PQ. The length of PQ is d. Then at PQ, power $d\mathbf{v}\mathbf{B}$ is induced and current is shed in the direction P → O → R → Q.

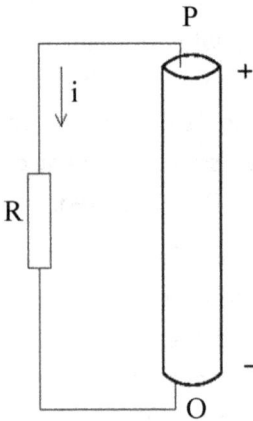

Fig. A2.6 Current due to inductive voltage.

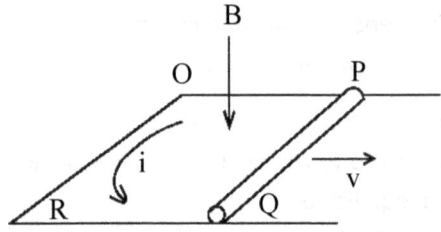

Fig. A2.7 Sliding movement of wire PQ. **B**: magnetic flux density; **v**: velocity; i: inductive current.

On the other hand, during time t, the area inside PORQ changes with $d\mathbf{v}t$, and the change of magnetic flux (the sum of magnetic flux density in PORQ) is given by $d\mathbf{v}t\mathbf{B}$. Therefore, considering inductive voltage to be $d\mathbf{v}\mathbf{B}$, the following relation is obtained:

Inductive voltage = change of magnetic flux per unit time.

This relation is a law of electromagnetic induction. Magnetism due to inductive current is in the inverse direction to **B**. Thus, it is found that in magnetic flux density **B**, current is induced so magnetism occurs in an inverse direction to **B**.

— ⟫⟪ — ⟫ ⟪ — ⟫⟪ —

2.8 Discovery of Laws of Electrochemical Decomposition

In the voltaic electric battery, several combinations of first conductors, such as zinc and copper, were stacked one on top of the other, and between each combination, a second conductor, such as an electrolyte, was inserted. Hence the battery was called a voltaic pile (Appendix 2.5). In 1800, Volta informed Banks, the president of the Royal Society, of this invention.

Banks, after receiving Volta's letter, informed William Nicholson, an English chemist, and Anthony Carlisle, a surgeon, about the invention of the voltaic pile. Nicholson and Carlisle fabricated a voltaic pile and utilized the battery as a source of electric power. As shown in Fig. 2.15, the ends of two wires, acting as electrodes, were connected to the terminals of the battery and their other ends were dipped in a vessel filled with water. As electricity from the battery passed through the wires, hydrogen accumulated at the cathode wire and oxygen accumulated at the anode wire, that is, electrochemical decomposition of water was verified. Nicholson and Carlisle were the first ones to verify the chemical reaction.

In 1811, Faraday verified the idea proposed Joseph Louis Gay-Lussac and Louis Jacques Thenard that the factor controlling the rate of decomposing of an electrolyte is not the thickness of

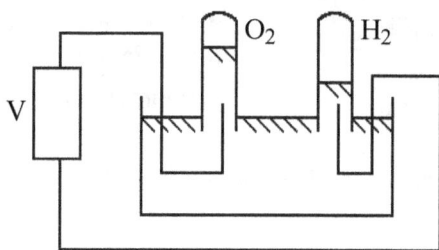

Fig. 2.15 Electrochemical decomposition of water. V: battery; O_2: oxygen gas; H_2: hydrogen gas.

the solution and the electrode but the intensity of the current passing through the solution. In other words, it was concluded that in electrochemical decomposition, the connection of the electrode with solution was not necessary and the process of electrochemical decomposition did not depend on the action of the electrode but on the intensity of the current passing through the solution.

—⟫⟪— ⟫ ⟪ —⟫⟪—

Appendix 2.5 Invention of the voltaic pile

In 1780, Luigi Galvani (Fig. A2.8), professor of anatomy at the University of Bologna, dissected a frog and brought a surgical knife in contact with a nerve of the frog's leg. The muscles of the frog's leg twitched (Yukawa & Tamura, 1955–1962). Galvani published his theory that the muscles of a frog produces electricity. This discovery had a great influence on physicists and physiologists.

Fig. A2.8 Luigi Galvani (1737–1798).

Volta repeated the experiment with the frog's leg. He set silver foils at two separate points on a nerve of a frog's leg and passed electricity using the silver foils as electrodes. The muscles of the frog's leg cramped. From this experiment, he concluded that a stream of electricity stimulated the nerve and as a secondary effect from the nerve, the muscles cramped.

He accomplished another experiment with a different method. He connected a wire to two silver foils placed at two separate points

Fig. A2.9 Alessandro Volta (1745–1827).
(Portrait made in the beginning of the nineteenth century.)

on the frog leg nerve but did not pass any electricity. The muscles cramped. When Volta connected the wire to the silver foils as electrodes, electricity formed due to closing of the circuit between nerve and wire, which stimulated the nerve, and as a secondary effect from the nerve, the muscle cramped. Using this fact, Volta challenged Galvani's theory.

Volta discovered that when two different metals, not silver foils, were set at two points on the nerve, the muscles cramped more intensively. He theorized that the presence of two different metals is the cause of a strong electric stream but dissection of a frog produces no electricity.

While experimenting with electric streams using different metals, Volta discovered that the contact of the first conductor, such as zinc, copper, or silver, and the second conductor, such as a wet conductor or a fluid, created an electric stream. He discovered that inserting a second conductor, such as matter dipped in salt-water (electrolyte), between two electrodes of the first conductor, such as zinc and copper, causes an electric stream. Making a stack of such pairs (where a second conductor was inserted between a pair of electrodes of the first conductor), a voltaic pile was invented (Fig. A2.10).

Zinc
Electrolyte
Copper

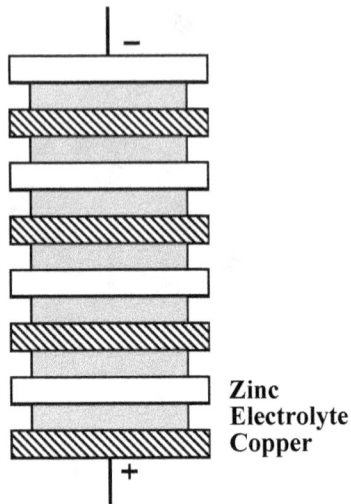

Fig. A2.10 A voltaic pile.

Utilizing a voltaic pile as a direct current power supplier, Davy, Faraday, and others performed electrochemical decomposition. Today, we have rechargeable batteries, such as lithium-ion batteries, capable of being repeatedly recharged after discharging. In contrast to batteries based on chemical reactions, there are batteries based on physical phenomena, such as solar batteries, which produce power using radiating semiconductors consisting of p-n junction (Shioyama, 2002).

Laws of electrochemical decomposition discovered by Faraday

In 1833, Faraday discovered the first law of electrochemical decomposition, that the quantity of products due to electrochemical decomposition is proportional to the absolute quantity of electricity passing through the solution (Explanation 2.6). Furthermore, measuring the mass of the constituents obtained by decomposition, he found that the quantities of the constituents produced at the cathode and the anode had a constant ratio. The quantity determining this ratio was called the "electric chemical equivalent." From measuring the chemical equivalent, Faraday discovered the second law of electrochemical decomposition, the electric chemical equivalent coincides with the chemical equivalent and the ratio of quantities of products on decomposition is equal to the ratio of their chemical equivalents (Explanation 2.6).

According to the second law, for example, the ratio of hydrogen and oxygen produced by the decomposition of water is the ratio of their chemical equivalents 1.008:8, though the molar ratio of hydrogen and oxygen is 1:1/2 and the ratio of their volumes is 2:1 (Explanation 2.6).

— ·»} {{· — ·»} {{· — ·»} {{· —

Explanation 2.6 Laws of electrochemical decomposition

Faraday's laws of electrochemical decomposition are explained quantitatively.

The first law: The quantity of the constituents produced at the electrodes by electrochemical decomposition is proportional to the quantity of electricity (ampere × second).

The second law: The quantity of constituents produced by the same amount of electricity is proportional to their chemical equivalents.

For example, the ratio of quantities of H_2 and O_2 produced by the decomposition of water is equal to the ratio of their chemical equivalents 1.008:8. Their molar ratio is 1:1/2. According to Avogadro's law, the ratio of their volumes is 2:1, that is, the volume of H_2 is double that of O_2.

The third law: The quantity of electricity required for the precipitation of 1 gram chemical equivalent = 96,500 coulomb = 1 faraday.

1 coulomb = quantity of electricity given by 1 ampere × 1 second.

1 gram chemical equivalent = quantity of the constituent expressed in gram with a numerical value of the chemical equivalent.

Avogadro's law: Equal volumes of all gases have the same number of molecules under the same temperature and pressure.

Explanation 2.7 Chemical equivalent

The ratio of quantities of constituents produced at the anode and the cathode by electrochemical decomposition is determined by the ratio of their chemical equivalents.

Atomic weight represents relative mass, and the mass of one atom of oxygen is 16. A gram atom of a substance is the quantity of the substance whose weight in grams is numerically equal to the atomic weight of the substance. For example, 1 gram atom of oxygen is 16 g. The quantity of molecules represents the relative mass of the molecules. The mass of one molecule of oxygen is 32. A gram molecule (mol) of a substance is the quantity of the substance whose weight in grams is numerically equal to the molecular weight of the substance. For example, 1 gram molecule of nitrogen (N) of 1 mol is 28.016 g.

In 1 gram molecule (mol) of a substance, there are
$6.02214076 \times 10^{23}$ (Avogadro's number) molecules.

Chemical equivalent is defined as the quantity of a substance that can combine with 0.5 atomic weight of oxygen or 1 atomic weight of hydrogen. For example, the chemical equivalent of oxygen is 8 (from H_2O), and the chemical equivalent of chlorine is 35.457 (from HCl).

When the atomic weight and valence are known, the chemical equivalent is given by atomic weight/valence, where valence is defined as the number of hydrogen atoms combining with one atom of the element. For example, the valence is 1 for Cl (HCl), 2 for O (H_2O), and 3 for N (NH_3). The valence of a constituent not combining with hydrogen is known by the ratio of the number combining with other elements with known valences.

— ·›〉〈‹· — ·›〉 〉‹· — ·›〉〈‹· —

Utilizing the precipitation of a metal in solution by decomposition, Davy successfully isolated sodium and potassium for the first time. Today, plating is performed using decomposition.

In 1834, Faraday continued research on electrochemical decomposition and introduced new academic terms, consulting linguists, in order to express precisely the phenomenon of electrochemical decomposition and make discussion among scientists easy. Some new academic terms he introduced were "electrode," "anode," "cathode," "electrolyte," "anion," "cation," and "ion." These terms are still in use today. A metal in solution is expressed by the term "anion."

2.9 Research on Dielectrics, Light and Magnetism, and Magnetic Substances

Research on dielectrics

When doing research on electrochemical decomposition, Faraday paid attention to the effect based on the law of electric conduction. An electrolyte remains conductive as a liquid but loses this property when it solidifies. For example, when water becomes ice, it cannot conduct current. When Faraday set platinum foils on both upper and lower surfaces of a block of ice and connected the foils to power, an electric charge was induced. When the ice was liquefied to water, it conducted current.

The phenomenon in which charge was induced on the surface of an insulator set between two metals (electrodes) connected to power terminals occurred by the action of polarized contiguous particles (dipoles) in all insulator medium. Faraday thought that the electric action between separate points occurred through the medium of matter. Thus, Faraday related charge induction on the surface of the insulator medium to the polarization of the insulator medium placed between the electrodes (Explanation 2.8). To perform research on the effect of polarization, he performed an experiment to understand how the charge induced on an insulator medium between two metals (electrodes) depended on the insulator medium mediating between the two metals.

Consequently, he found that for the same voltage, the charges induced on surfaces of insulator mediums were different for different insulator mediums. Faraday called the relation between charge and voltage "dielectric capacitance." This is related to the dielectric constant. In November 1837, he proposed to call the insulator medium a "dielectrics."

—◦}} {{◦— ◦}} {{◦ —◦}} {{◦—

Explanation 2.8 Dielectric constant

When, as shown in Fig. E2.10, two particles with electric charges q_1 and q_2 are placed distance r apart, the static electric force F influencing the particles is proportional to the product of q_1 and q_2 and inversely proportional to the square of the distance r. If both charges are positive or negative, then the static electric force is repulsive, and if one charge is negative and one is positive, the static electric force is attractive. This is called Coulomb's law (Yukawa & Tamura, 1955–1962).

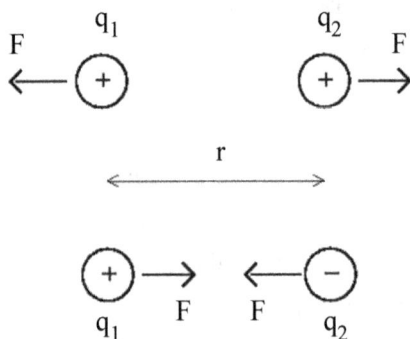

Fig. E2.10 Coulomb's law.

Next, we consider molecules in dielectrics (insulators). If in a molecule, the center of gravity of the positive charge does not coincide with that of the negative charge and the two centers are separated, the molecule is called a polar molecule. On the other hand, if both centers of gravity coincide, then the molecule is called a nonpolar molecule.

When (Fig. E2.11) polar molecules are placed between two metal bars, one negatively charged and one positively charged, the polar molecules rotate in the direction of the static electricity. On the other hand, when a non-polar molecule is placed between the two metal bars, the positive and negative centers of gravity move in the direction of the static electricity and the molecule becomes a dipole. That is, both polar and nonpolar molecules are directed by the static electricity as dipoles (Fig. E2.11). Consequently, according to the direction of the dipoles in the static electricity, on the upper surface of the dielectrics placed within the static electricity, positive charges appear and on the lower surface, negative charges appear (Fig. E2.12). This phenomenon is called "dielectric polarization."

When the voltage between upper and lower electrodes is V (Fig. E2.13), the charge Q induced on the surface of the dielectrics placed between the two electrodes is proportional to V. The proportional coefficient C is called "capacitance." As shown in Fig. E2.13, when the area of the electrodes of

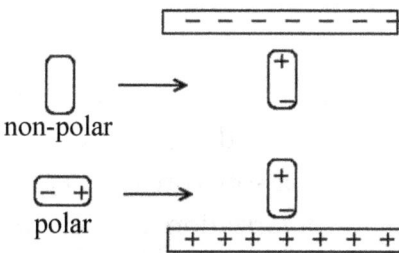

Fig. E2.11 Dielectric polarization.

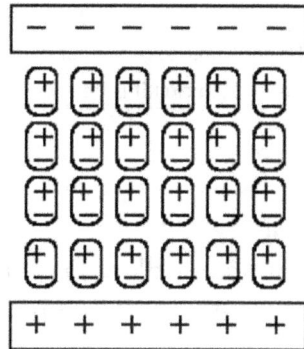

Fig. E2.12 Molecules in static electricity.

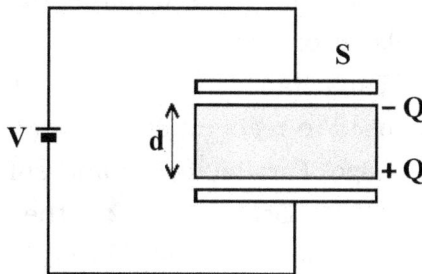

Fig. E2.13 A condenser. V: voltage of power supply; S: area of the electrode; d: distance between electrodes; Q: charge induced on the surface of dielectrics.

the condenser (also called the capacitor) is S and the distance between the electrodes is d, the capacitance C is proportional to the ratio S/d. This proportional coefficient is called the "dielectric constant."

The condenser is an important electronic part consisting in an electronic circuit together with a resistor and a coil. Using dielectrics with large dielectric constants, we can fabricate a condenser with a large capacitance.

A condenser does not allow direct current to pass through it, only alternating current. Therefore, in an electronic circuit, a condenser is utilized as a high-pass filter for the purpose of allowing high-frequency alternating current. A high-pass filter allows alternating current with a higher frequency than the threshold to pass through it. The threshold is called the "cut-off frequency" and is inversely proportional to the capacitance of the condenser. On the other hand, when direct current is treated in an electronic circuit, a smooth circuit is used in order to remove any unnecessary ripples in the signal. In a smooth circuit, a condenser is used in order to remove the ripples by earthing. When the capacitance of the condenser in the smooth circuit is large, the function of the smooth circuit is better.

—·》《·— ·》 《· —·》《·—

Particles in dielectrics can be compared to a system of small magnetic needles. Faraday thought that the polarization of dielectrics is similar to the polarization of soft iron in a magnetic field—that is, he thought that the polarization of dielectrics could be explained by an electric field. Faraday introduced the idea of an electric line of force. Thus, the concept of electric field was born, which Maxwell used to theorize electromagnetic phenomena (Explanation 2.5).

As the unit of capacitance of a condenser (also called capacitor), farad (F) is used in recognition of Faraday's contribution.

In November 1839, Faraday fell ill and suffered from dizziness and headache. In December 1840, the Royal Institution exempted him from duty till he felt better. In 1841, he went to Switzerland for three months to rest and recuperate but he never did recover completely.

Research on light and magnetism

As shown in Fig. 2.16, light is a wave vibrating in a direction perpendicular to the direction it moves—it vibrates in any direction in the plane perpendicular to the line it moves. But when it passes through a special crystal, it becomes polarized, with a constant direction of vibration.

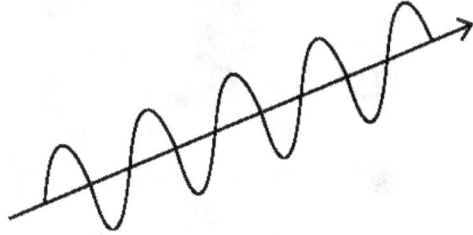

Fig. 2.16 Vibration of light.

In August 1845, William Thomson, Faraday's friend and a mathematical physicist sent a letter to Faraday asking what effect a transparent dielectrics would have on polarized light (James, 2010, p. 79). To answer the question, Faraday tried an experiment.

Believing that magnetic and electric action were correctly explained by the concept of line of force, Faraday tried to understand magnetic property using magnetic lines of force. To examine the relation between light and magnetism, he used a heavy glass as the polarizer, set a powerful electromagnet near the heavy glass, and allowed light to pass through the heavy glass (Fig. 2.17). The polarization plane rotated. The rotation of the polarized plane of light due to the magnetic field was called the magneto-optical effect, or Faraday's effect. It was discovered on September 13, 1845.

There is a method for measuring current in a high-voltage power-transmission line with noncontact safety—an example of application Faraday's effect. The current is measured utilizing the change of rotation angle of the polarization plane in Faraday's effect depending on the intensity of the magnetic field induced by the current passing through the power-transmission line.

Fig. 2.17 An electromagnet, currently in Faraday Museum, used by Faraday in his research on light and magnetism. (Photograph taken by Dr. K. Matsuda in September 2017.)

In 1862, Faraday performed an experiment on change of wavelength of light passing through a strong magnetic field. However, because the sensitivity of the apparatus was insufficient, he was not successful. The phenomenon Faraday expected to see was discovered by Pieter Zeeman in 1896 and was called the "Zeeman effect" (a phenomenon where the wavelength of light splits into multiple wavelengths). The fact that the phenomenon expected by Faraday was discovered about 30 years later indicates that Faraday carried out research on the phenomenon strictly on the basis of acute intuition.

Research on magnetic substances

Faraday performed research on the magnetic behavior of substances magnetized by a magnetic field. He would hang a piece of the substance in a strong magnetic field so that it could move freely about. Substances like glass moved in a direction perpendicular to the direction of the magnetic line of force. But paramagnetic substances, such as aluminum, moved in the direction of the magnetic line of force. Substances like glass behaved the way they did because magnetization inverse to the magnetic field occurred. Faraday called the magnetic property of substances different from paramagnetism "diamagnetism" in 1845.

2.10 Social Contribution by Faraday

Lighting

Lighting is considered to be Faraday's social contribution. In 1836, Faraday joined as advisor to Trinity House. He continued this responsibility till 1865. Trinity House was a station responsible for the safety of coast voyage, and changing old lighting was important work. Utilizing his ability, he planned reducing fuel consumption and improving the efficiency of lighthouse lighting.

In 1840, Faraday invented a new chimney for oil lamps. This chimney efficiently removed the gas created because of oil

Fig. 2.18 Faraday.
(Portrait made in 1842).

combustion and decreased the cloudiness of the lamp glass, improving the lighting.

This chimney was applied to not only the lighthouse on the English coast but also the library at the Buckingham Palace. A leading newspaper reported that the oil lamp lighted Princess Helena's baptism. Because Faraday was not interested in a patent, in 1842, he transferred the patent of his chimney to his brother.

Fig. 2.19 Bust of Faraday by Matthew Noble, 1854, currently in Faraday Museum. (Photograph taken by Dr. K. Matsuda in September 2017).

In 1854, William Watson, from Trinity House, requested Faraday to test the electric lighting system he had invented (Fig. 2.20) in 1852. The system utilized an arc discharge that occurred between the carbon electrodes using voltage supplied by the battery.

Faraday summarized the test results in a report of 4200 words and reported the following regarding the battery in terms of its use in a lighthouse:

- The problem of collection of the chemical substances produced by the battery should be solved.

Fig. 2.20 William Watson (1715–1787).
(Printed in 1784).

- A lot of room is necessary for the battery.
- A room that could accommodate three persons, for maintenance of the battery, was necessary.
- The brightness of the arc discharge varied with time.
- At the present stage, it was difficult to find persons for maintaining this electric lighting system.

He concluded that the electric lighting system could not be used in a lighthouse.

In 1857, Frederick Holmes proposed another electric lighting system using arc discharge. In Watson's system, a battery was used, but in Holmes's system, a generator driven by a steam engine based on electromagnetic induction, discovered by Faraday, was used. By Faraday's positive recommendation, Trinity House approved a budget for the system's test run. On December 8, 1857, in the presence of Faraday, the electric lighting system lighted the English Channel for the first time. But afterward, because of technical problems, Holmes's system was given up. The use of the oil lamp was continued in the light house until in 1920 it lit with incandescent light, produced by Joseph Swan.

Other social contribution

From 1829 to 1851, Faraday served as professor of chemistry at the English Military Academy.

In 1840, Faraday was elected as an elder of a Sandemanian church. The elder preached, baptized infants, and presided at the Love Feast. In 1860, he was again elected as the elder and served in the role till 1864. He was not eloquent. But due to his personality, he continued as an elder for a long time. Before he was elected as the elder, he performed work supporting an orphanage (from about 1832). For Faraday, church, family, and work were related and his living and work (science) were understood only by his religious belief and practice.

On September 28, 1844, in Haswell Colliery, an explosion occurred, killing 95 people, including three young boys. Prime Minister Robert Peel requested Faraday and Charles Lyell to investigate the accident. On October 8, 1844, they went to Haswell to investigate and reported that it was important to improve the ventilation in collieries in order to prevent them from being filled by firedamp. In October, before they left Haswell for London, they contributed to establishing a fund for supporting the widows and orphans of the explosion.

In the war of Britain and France against Russia (the Crimean War), to prevent the Russian empire from expanding into western Europe due to the decline of the Turkish empire, Britain allied with France and attacked Cronstadt on the Baltic Sea and Sevastopol on the Crimean Peninsula, in order to inflict damage on trade. Faraday was often asked for technical advice in secret by the Royal Navy.

In 1854, Thomas Cochrane proposed the use of sulfur-filled fire ships to attack Cronstadt. Faraday, whose opinion was sought on the proposition, analyzed the poison gas chemically and insisted that it was necessary to know the situation if sulfur-filled fire ships were to be used. He reported that a chemical weapon should not be used. On the basis of Faraday's report, James Graham, the minister of navy, dismissed Cochrane's proposition. Thus, always taking into account the influence on citizens, Faraday gave cautious counsel.

2.11 A Grace-and-Favor House at Hampton Court Offered by Queen Victoria to Faraday and His Wife

As mentioned earlier, since November 1839, Faraday had suffered from poor health. His wife was also not well and had difficulty walking.

Queen Victoria heard from Prince Albert that Mr. and Mrs. Faraday had been living in the attic of the Royal Institution for 37 years and because of poor health had difficulty negotiating the stairs. She offered them, for life, the use a grace-and-favor house at Hampton Court, facing the upper stream of the Thames (Fig. 2.21). From then on till their deaths, they lived in the graceful residence.

Fig. 2.21 Hampton Court Palace. (Photograph taken from https://www.britainexpress.com/attractions.htm? attraction=169).

Fig. 2.22 Photograph of Faraday taken in the 1860s.

Faraday was not interested in honor. Therefore, he declined the position of president of the Royal Society twice. He also declined the presidency of the Royal Institution. He had no children and always feared that his wife would live in solitude after his death. On August 25, 1867, he passed away quietly sitting in his armchair. He was 76 years of age. He was buried at High-Gate Cemetery.

Faraday was born of a blacksmith and his wife. At 13 years of age, he gave up regular schooling and started his life as a newspaper-cumerrand boy at a bookshop and stationers. Yearning to be a scientific practitioner, he endeavored self-studying and got a position in the Royal Institution of Great Britain. Having accomplished many epoch-making discoveries, such as electromagnetic induction and magneto-optical effect, Faraday was given the honor of being called "Prince of Science." He was innately kind and widely admired. The respect people gave him is larger than the honors he received. He believed that humans could have a noble mind by researching science and

Fig. 2.23 Photograph of Faraday taken in a later part of his life.

contributed to the development of mankind by accomplishing great works. It is not too much to say that his teacher Davy's biggest discovery was Faraday.

Six years after Faraday passed away, 50,000 miles of cable was laid for telegraphy. The telegraph-cable-laying ship was named *The Faraday* with the permission of his wife. Simultaneously, by naming the unit of capacitance "farad," the industry of electricity paid Faraday the highest possible compliment as one of nineteenth century's most celebrated natural philosophers.

References

Bowers, B., & Tamura, Y. (trans. 1978), ファラデーと電磁気. Tokyo Tosho, 東京図書. (1974). *Michael Faraday and electricity*. Wayland Publishers Ltd.

Davy, H. (1816). On the fire-damp of coal mines, and on methods of lighting the mines so as to prevent its explosion. *Philosophical Transactions of the Royal Society*, **106**, 1–22.

Faraday, M. (1832), Experimental researches in electricity, *Philosophical Transactions of the Royal Society*, **122**, 125–162.

Faraday, M., & Takeuchi, Y. (trans. 2010), ロウソクの科学. Iwanami Shoten, 岩波書店. (1861). *The chemical history of a candle.*

James, F. A. J. L. (Ed.). (1991). *The correspondence of Michael Faraday: Volume 1: 1811-1831.* London: The Institution of Electrical Engineers.

James, F. A. J. L. (2010). *Michael Faraday: A very short introduction.* Oxford: Oxford University Press.

Kekule, F. A. (1865). Sur la constitution des substances aroma- tiques. *Bulletin de la Societe Chimique de Paris,* 3, 98-110.

Mendelssohn, K., & Ooshima, K. (trans. 1971). Kodansha, 講談社. (1966). *The quest for absolute zero.* Weidenfeld and Nicolson.

Shioyama, T. (2002). *Principle and application of sensor,* センサの原理 と応用. Morikita Shuppan, 森北出版.

Sootin, H., Koide, S., & Tamura, Y. (trans. 1976), ファラデーの生涯. Tokyo Tosho, 東京図書. (1954). *Life of Michael Faraday.* A division of Simon & Schuster Inc. New York.

Tyndall, J. (2002). *Faraday as a discoverer.* McLean, VA: IndyPublish. com. (Original work published in 1868, New York: D. Appleton).

Yukawa, H., & Tamura, S. (1955-1962). *Accepted theory of physics* (Vols. I–III), 物理学通論, 上・中・下. Tokyo: Taimeidou, 大明堂.

Chapter 3
Albert Einstein

A lbert Einstein formulated the relativity theory and derived the new concept of time-space, introducing a revolution in physics. The relativity theory and quantum mechanics are the two greatest theories of modern physics. The relativity theory has been a powerful guideline in the development of cosmology. Among the scientists at the beginning of the twentieth century, no one made more contribution to epoch-making progress in physics than Einstein.

3.1 Upbringing

Birth of Einstein

On March 14, 1879, Einstein was born at Ulm, in southern Germany (Fig. 3.2). His parents were Jews but they were not enthusiastic Jews and did not follow Jewish customs. His father, Hermann Einstein, was a calm and kind man loved by all. He liked literature and read Schiller and Heine to family members. He had a small electrical construction business but was in financial difficulties. When Einstein was one year old, the family relocated to Munich. On November 18, 1881, his sister Maria (always called Maja) was born (Fig. 3.1). Maja was to become her brother's most intimate soulmate (Isaacson, 2017, p. 11).

In Munich, his father, along with his uncle Yakob, started an enterprise that offered facilities related to gas and tap water. In 1885, they started an electrotechnical factory to produce

Fig. 3.1 Einstein and his sister Maria.
(Photograph taken in 1885).

Fig. 3.2 Present Ulm (in the vicinity of the Ulm Cathedral).

dynamos, arc lamps, and electrical measuring equipment for municipal electric power stations and lighting systems. The fund was invested by Einstein's maternal grandfather. Einstein and his sister liked their life at home, which had a large garden with large trees.

A late speaker

Einstein's head was unusually large for a baby's head. He liked being alone. He was not interested in games and toys, unlike other children. The child might have seemed backward for his age because he began speaking unusually late. Whenever he had something to say, he would try it out on himself, whispering it softly until it sounded good enough to say out loud. His younger sister recalled that irrespective of how routine the sentence was, he repeated it to himself softly, moving his lips (Isaacson, 2017, p. 8). His mother used to worry initially that there was something wrong with him. But as the reason became clear, her anxiety abated. The reason was that Einstein's thought process was internally profound, and before uttering any word, he would think it through thoroughly. Hence, he was taciturn and a very quiet boy. At the age of five, when he was sick in bed, his father gave him a small compass to keep him occupied. The compass really impressed him. "Something deeply hidden had to be behind things," said he afterward about this experience of his early in his life (Pais, 1982, p. 37).

A competent pianist, his mother, Pauline, decided to introduce her children to musical education. At the age of six, Einstein took violin instruction and his sister learned the piano. He disliked the repetitive nature of practice and was not interested in the violin. But once he heard a sonata of Mozart's, music became important to him and in time, he became a skillful violin player. Music was not only a diversion but also an important aid. It helped him to think (Isaacson, 2017, p. 14)—that is, when he faced a challenge in his work, playing the violin helped him resolve the issue (Isaacson, 2017, p. 14).

Entering public school

At the age of six, Einstein enrolled in a *volksschule*, a public school. He was a superior student and a patient one at that and solved mathematical problems with confidence. Because the games he liked necessitated patience and persistence, he did not play with his classmates and was basically a gentle boy.

When Einstein was seven years old, his uncle Yakob started to teach him algebra. Yakob would give him increasingly difficult problems to solve and Einstein would always solve them, taking delight in the task (Isaacson, 2017, p. 17). In October 1888, He entered the Luitpold-Gymnasium, a secondary school in Munich, Germany. He was always the most superior of students and was especially good in mathematics and Latin. However, he was not satisfied with the school because in the school there were authoritarian teachers and servile students and rote learning was the norm.

Max Talmud, a medical student with little money, came for dinner every Thursday night. After dinner, he discussed science and philosophy with Einstein. He presented him a scientific book and was an important educational influence on Einstein.

In 1894, his father's business failed and they planned to move the factory to Italy according to recommendation of Italian Signor Garrone, a comrade of the enterprise. The family relocated to Milan, but Einstein stayed on in Munich to complete his schooling. In 1895, after the factory was completed, the family again relocated to Pavia.

Fig. 3.3 Einstein at the age of 14. (Photograph taken in 1893.)

While staying in Munich, Einstein was depressing and missed his family. He disliked school. Without consulting his parents, he decided to go to Italy. Requesting the doctor to prepare a medical certificate showing Einstein to have poor physical health, he took a leave of absence from the Luitpold-Gymnasium and in 1895, he relocated to Pavia. To his parents, who were astonished by his sudden arrival, he informed that later, he will do self-studies and try the entrance examination for ETH (Eidgenossische Technische Hochschule Zurich: ETH Zurich) (Pais, 1982, p. 40). He disliked the authoritarian education in the Luitpold-Gymnasium and ultimately left the school halfway. Furthermore, he decided to give up his German citizenship. He started his new life in Italy, where the landscape and the arts impressed him profoundly (Pais, 1982, p. 40). In Italy, the quiet boy suddenly became lively and a talkative youth.

3.2 Eidgenossische Technische Hochschule Zurich: ETH Zurich

Entrance examination for ETH

In October, at the age of 16, Einstein took the entrance examination for ETH in order to study electrical engineering, but he failed in the examination. As previously decided, he gave up his German citizenship. Afterward, for some years, he would not get any citizenship.

Fig. 3.4 ETH Zurich. (Photograph taken from https://www.reisen-experten.de/reise-news/zuerich-soll-als-smart-destination-positioniert-werden/.)

To prepare again for the entrance examination of ETH (Fig. 3.4), he went to the cantonal school in Aarau, a predominantly German-speaking area on the Swiss plateau. He boarded at the house of Winteler, a teacher at the school. The Wintelers were good people. There was free atmosphere in the school, and Einstein came to respect the teachers and could enjoy school life. The scars of the failure he had faced in the entrance examination faded away. He thought that if he passed the entrance examination, he would go to Zurich. For four years there, he would study mathematics and physics. In the future, he would be a teacher of science, especially the theoretical field. Abstract mathematical thinking fitted his temperament (Pais, 1982, p. 40).

In 1896, the factory his father and uncle ran went bankrupt. The uncle found work in a large company, but his father decided to start a new factory. Einstein warned his father against starting a new enterprise and visited his uncle to request him not to support his father in his endeavor. Einstein sent to his sister a letter telling her that his parents' repeated misfortune weighed heavily on his mind and that as an adult he felt a burden on the family and it anguished him not to be able to support his parents (Pais, 1982, p. 41). Two years later, his father found work in an electric company and Einstein's depression ended.

Studying by himself

On October 29, 1896, at the age of 17, Einstein passed the entrance examination for ETH. Upon satisfactory completion of the four-year curriculum, he would qualify as a *fachlehrer*, a specialized teacher in mathematics and physics at a high school. He did not attend lectures and studied by himself by going through the research studies of Gustav Kirchhoff, Heinrich Rudolph Hertz, and Hermann Ludwig Ferdinand von Helmholtz and the electromagnetic theory of James Clerk Maxwell. He considered Hermann Minkowski to be a superior teacher of mathematics but did not attend his lectures.

Not relying only on the lectures of the university, he expanded the view of his field of academy. He read papers by Hendrik Antoon Lorentz (Fig. 3.15) and Ludwig Eduard Boltzmann. In Zurich, he got acquainted with many friends and lived a pleasant student life. Four years later, in 1900, at the age of 21, Einstein became qualified as a *fachlehrer*. The maximum one could get in the examination was 6.0, and his final grades were 5.0 for theoretical physics, 5.0 for experimental physics, 5.0 for astronomy, 5.5 for function theory, and 4.5 for a short paper on thermal conduction (Pais, 1982, pp. 44, 45). Because he did not attend lectures, he prepared for the examination by borrowing notebooks from Marcel Grossmann. In the same year, Max Karl Ernst Ludwig Planck (Fig. A3.1) proposed the concept of energy quantum (Appendix 3.1).

3.3 The Patent Office in Bern

Unsuccessful application for a post at the university

Because Einstein graduated university with superior grades, he reasonably expected to obtain a post at the university. Einstein, who had left the Luitpold-Gymnasium halfway disliking the principle of education, continued in his youth in a similar vein. At that time, it was common to address a professor with the honorific "Herr Professor" but while he attended ETH, he did not address a professor with the honorific. He was considered to have a sassy attitude. Einstein's experimental plan on the Earth's movement against the aether was not permitted by Heinrich Friedrich Weber, who was the professor in charge of Einstein. One day, Weber told Einstein that he is a very smart boy but has a serious defect—of not listening to a person (Pais, 1982, p. 44). The term "smart" can mean "clever" as well as "cunning." As a result, Einstein's enthusiasm and fascination for experiment gradually faded. Although the post of an assistant was vacant, the professor did not intend to offer the post to Einstein. He wrote to Heike Kamerlingh Onnes and Friedrich Wilhelm Ostwald in Leiden and Leipzig, respectively, letters applying for a post, but his applications were unsuccessful. Three other students in the same class as Einstein immediately obtained positions as assistants at ETH.

In 1901, he got Swiss nationality. After giving up on getting a post at the university, on May 19, he joined as a substitute teacher at an industrial high school in Winterthur, aware that as a teacher he could carry on his efforts in science. He wrote to Grossmann a letter describing how he researched on kinetic gas theory and on the relative motion of body against aether (Pais, 1982, p. 46). On September 15, he got a temporary position at a private school in Schaffhausen.

In November of the year, to get a doctoral degree, he submitted a thesis on kinetic gas theory to the University of Zurich. In those days, ETH did not confer doctoral degrees. However, his thesis was not accepted as a doctoral thesis. This was his last failure.

Grossmann's kindliness

Grossmann (Fig. 3.7) told his father about Einstein's difficulty in finding employment. His father informed Friedrich Haller, president of the Patent Office in Bern, about this. On December 11, 1901, one post fell vacant in the Patent Office. Immediately, Haller interviewed Einstein and guaranteed him the post.

In February 1902, Einstein resigned from the public school in Schaffhausen and relocated to Bern. His living depended on his income as a private teacher and a little money he received from home. Students he taught as a private teacher, Maurice Solovine and Konrad Habicht (Fig. 3.5), became his close friends, and they regularly met to discuss philosophy, physics, and literature over a simple meal. They called the meetings "Akademie Olympia."

On June 16, 1902, he joined the Patent Office. To begin with, he was employed as the third type of technical staff in a temporary position, and on September 16, 1904 (Fig. 3.6), he was employed as regular staff. On April 1, 1906, he became the second type of technical staff. His work there was to examine whether a

Fig. 3.5 Habicht, Solovine, and Einstein. (Photograph taken in 1903).

Fig. 3.6 Einstein in 1904.

patent applied for was based on a scientific principle or not.

In the Patent Office, during daytime break, to avoid wasting time, he avoided contact with people and would be lost in thought in his spare time. People took him to be an ordinary public official. No one would have thought then that he would go on to accomplish great discoveries.

After the Patent Office's working hours, at home, Einstein continued his theoretical research. Because the Patent Office had no academic space, unlike the university, he carried on research at home. The process of accumulating research work little by little gave him a sense of fulfillment. Though he could not get a doctoral degree in 1901, he did further research and submitted a thesis entitled *Eine Neue Bestimmung der Moleküldimensionen* to the University of Zurich in 1905 and got a doctoral degree. He dedicated the doctoral dissertation to Grossmann. Grossmann also got a doctoral degree at the same university.

Private life

Before Einstein relocated to Bern, he wanted to marry Mileva Marity, with whom he used to discuss science at ETH. She was

Fig. 3.7 Marcel Grossmann (1878–1936).

born in Titel, South Hungary, in 1875 and had a Greek Catholic background. Einstein's mother did not like her. Therefore, his parents strongly opposed the marriage. His father got a fatal heart disease. When Einstein visited his father, his father finally consented to the marriage. His father passed away on October 10, 1902. On January 6, 1903, Einstein married. On May 14, 1904, his son Hans Albert was born.

3.4 Publication of Three Papers

Three papers in *Annalen der Physik*

The research results Einstein accomplished at home while working at the Patent Office were published as three papers in *Annalen der Physik* in Germany. The first paper was on the photoelectric effect, entitled *Uber Einen die Erzeugung und Verwandlung des Lichtes Betreffenden Heuristischen Gesichtspunkt*; the second paper was on the special relativistic theory, entitled *Zur Elektrodynamik Bewegter Korper*; and the third paper was on Brownian motion, entitled *Uber Die von Der Molekularkinetischen*

Theorie der Warme Geforderte Bewegung von in Ruhenden Flussigkeiten Suspendierten Teilchen. All the papers proved to be revolutionary in the field of physics.

Energy quantum

The first paper verified the concept of energy quantum proposed by Planck (Appendix 3.1). The concept of energy quantum insisted that electromagnetic energy in heat radiation, such as black-body radiation, was integer times energy quantum and it was discrete (Fig. 3.8). The concept was incompatible with classical physics. According to the electromagnetic theory by Maxwell and thermodynamics, electromagnetic energy was considered to consist of waves and it was assumed to be continuous, not discrete.

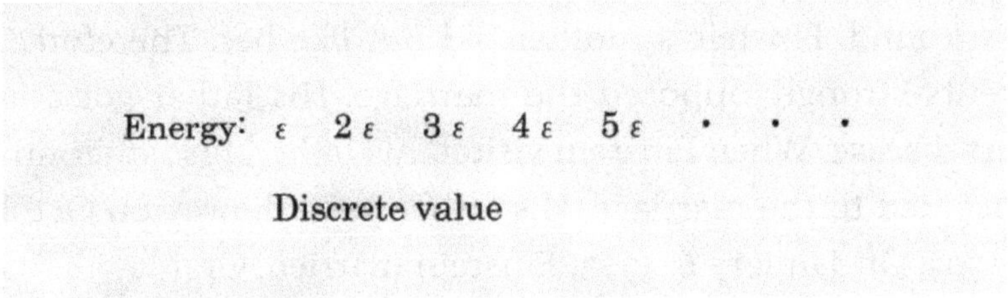

Energy: ε 2ε 3ε 4ε 5ε ·   ·   ·

Discrete value

Fig. 3.8 Energy quantum. ε: energy quantum.

However, when introducing the concept of energy quantum, Einstein called light with an energy quantum as a "photon" and he was successful in theoretically elucidating the phenomenon of the photoelectric effect, where an electron was emitted from a piece of metal irradiated by light. By this, the concept of energy quantum proposed by Planck was no longer a hypothesis but a truth. Instead of Newtonian mechanics, which could not be applied to the microscopic world, such as atoms, quantum mechanics based on the concept of quantum was developed at the beginning of the twentieth century. In 1922, Einstein was awarded the Nobel Prize in Physics for elucidating the photoelectric effect (Section 3.12 in this chapter). The photo-electric effect is applied to photosensors, such as photoelectric tubes and photoelectron multipliers (Shioyama, 2002).

Einstein's second paper was on special relativistic theory. This paper brought about a revolution in physics. The new theory will be described in Section 3.6.

—⟫⟪— ⟫ ⟪ —⟫⟪—

Appendix 3.1 Energy quantum

In 1900, when Planck derived Planck's formula, which theoretically solved heat radiation, which could not be explained by classical physics, he proposed the hypothesis that energy could not take a continuous value but was integer times inseparable unit of quantum. The energy quantum was related to frequency v and was given by hv, where h is Planck's constant. When n is a nonnegative integer, energy is given by the product of n and quantum. The concept of quantum at the beginning of the twentieth century led to new quantum mechanics, as mentioned in Appendix 1.3.

Fig. A3.1 Max Karl Ernst Ludwig Planck (1858–1947). (Photograph taken in 1890.)

—⟫⟪— ⟫ ⟪ —⟫⟪—

Brownian motion

Einstein's third paper, based on Boltzmann statistics, proved the existence of atoms. Maxwell and Boltzmann had given the kinetic gas theory, which assumed that a gas consisted of many atoms or molecules and derived its physical states, such as the pressure and thermal energy of the gas. Boltzmann extended the kinetic gas theory and developed Boltzmann statistics. But the question remained, does an atom exist? There was only one phenomenon that estimated the existence of an atom, the phenomenon discovered by Robert Brown on observing particles of pollen. The phenomenon was Brownian motion, where small particles suspended in a liquid were observed to jiggle around, and it was published in 1828.

Fig. 3.9 Brownian motion.

In his third paper, Einstein, considering that Brownian motion was brought about because of millions of random collisions due to atoms or molecules moving in all direction, theoretically derived the trajectory of a particle of pollen on the basis of Boltzmann statistics and indicated that the theoretical results coincided with the measured results in Brownian motion (Fig. 3.9). By this, the existence of an atom was verified.

—◈〉〈◈— ◈〉 〈◈ —◈〉〈◈—

Appendix 3.2 Planck's formula

Planck considered that black-body radiation consisted of oscillators with an electromagnetic vibration (a mechanical system with a sinusoidal vibration was called an oscillator) (Tomonaga, 1952). Expressing the frequency of the oscillator with an electromagnetic vibration by v, the number $Z(v)$ of oscillators with a frequency v per unit volume was described as $(8\pi c^3)/v^2$, taking into consideration the constraint for the wavelength of a stationary wave in an oscillator. Planck thought that the energy E of an oscillator was nonnegative times energy quantum ε and could have only discrete values. The average $<E>$ of E was given by $\varepsilon/[\exp(\varepsilon/k_BT) - 1]$ using the exponential function $\exp(\cdot)$, where T denotes the absolute temperature of the black body and k_B denotes the Boltzmann constant, which is given by the gas constant R divided by Avogadro's number.

Because the energy quantum ε was expressed by hv, the intensity of the black-body radiation $U(v)$, which was the product of the number of oscillators $Z(v)$ and the average of energy $<E>$, was given by Planck's formula (Fig. A3.2)

$$(8\pi h/c^3)v^3/[\exp\{hv/k_BT\} - 1].$$

Because the numerator of $U(v)$ is proportional to the cube of v, $U(v)$ increases with frequency. Because the denominator of $U(v)$ exponentially increases with frequency and the increasing rate becomes more than the increasing rate of the numerator, $U(v)$ decreases at some frequency. Hence, $U(v)$ has a maximum value at some frequency. For a black body with a high temperature T, the increasing rate of the exponential function of the denominator is smaller than that at a low temperature T, and the frequency at which $U(v)$ has a maximum value shifts to the higher side, according to Wien's displacement law. The formula coincided with experimental data of the relation of the intensity of black-body radiation to frequency. Planck's success in deriving the formula was because he introduced the concept of energy quantum, in which energy could take only discrete values of integer times quantum.

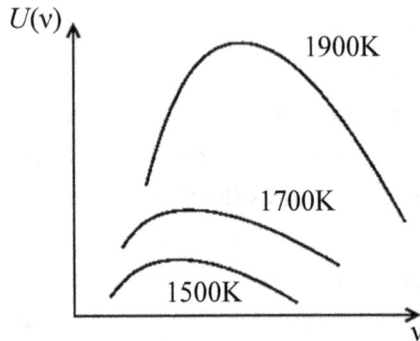

Fig. A3.2 Relation between the intensity of black-body radiation and the frequency of an electromagnetic wave. $U(v)$: intensity of black-body radiation; v: frequency. The frequency at which intensity takes the maximum value shifts to a higher side for a high-temperature black body.

—⟫⟪— ⟫ ⟪ —⟫⟪—

3.5 Historical Background of the Special Relativistic Theory

Aether

Before Einstein formulated the special relativistic theory, he researched the historical background of physics in detail. He summarized the problems and contradictions left. In the nineteenth century, Maxwell and others considered that the fundamental equations of electromagnetic theory held true only in inertial frames at rest against aether (Møller, Nagata & Ito, 1959; Yukawa & Tamura, 1955–1962). The inertial frame is a coordinate system where the coordinate axes cross at a right angle without distortion, moving at a constant velocity linearly, and Newton's first law, the law of inertia, holds true. In the nineteenth century, it was considered that aether penetrated every matter and

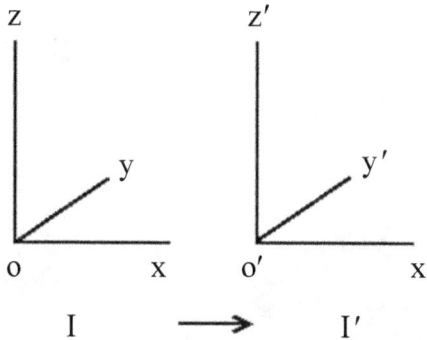

Fig. 3.10 Inertial frame I′ moving at a constant velocity in
direction *x* against inertial frame I.

vacuum and was the medium bearing every optical and electromagnetic phenomenon (Hawking & Sato, 2001; Pais, 1982).

But the question was whether aether existed or not. In order
to get an answer to this question, many scientists (mentioned
below) performed experiments researching optical phenomena
in inertial frames moving against aether. Using Galilean transformation, which was coordinate transformation, between inertial
frame I (at rest against aether) and inertial frame I′ (moving
against aether), experiments inspecting expected results were
performed. When inertial frame I′ moves at a velocity **v** in
direction *x* against inertial frame I, as shown in Fig. 3.10, the
velocity of light in inertial frame I′ in direction *x* is smaller than
that of the light in inertial frame I with velocity difference v.
Consequently, the velocity of light is different when light is
observed in different inertial frames.

Optical experiment by Michelson and Morley

Armand Hippolyte Louis Fizeau (Fig. 3.11) performed an experiment in 1859, and Jean Bernard Leon Foucault (Fig. 3.12)
performed an experiment in 1865. When the velocity of light
is expressed as **c**, the method measuring the square of **v/c** was
tried by Albert Abraham Michelson (Fig. 3.13) for the first time

Fig. 3.11 Armand Hippolyte Louis Fizau (1819–1896).

Fig. 3.12 Jean Bernard Leon Foucault (1819–1868).

in 1881. In 1887, six years later, the method was improved by Michelson and Edward William Morley (Fig. 3.14) (Appendix 3.3), as follows:

- They set up the equipment on a grand stone (1.5 m along every side and 0.3 m in thickness) that floated in a stratum of mercury so all equipment could rotate without distortion and not be influenced by vibration.
- They set many reflective mirrors that repeatedly reflected light, and the light pass length became 10 times the original because the estimated displacement of the interference fringe was very small in the inertial frames moving against aether and the displacement of the interference fringe was proportional to the light pass length (Appendix 3.3).
- They covered all optical equipment with wood covering to protect it from air stream and temperature change.

Thus, they fabricated the most precise experimental equipment by devising such grand improvements.

The displacement of the interference fringe was way smaller than estimated (Appendix 3.3). The estimated displacement of the interference fringe occurred assuming that the velocity of light in the inertial frame at rest against aether was different from that in the inertial frame moving against aether. Logically, if the proposition "if A holds true, then B holds true" is true, then the contraposition "if B does not hold true, then A does not hold true" is also true. In this case, A is "the velocity of light is different in different inertial frames" and B is "displacement of interference fringe occurs." Because the contraposition is true, it is concluded that "if there is no displacement of the interference fringe, then the velocity of light is the same in different inertial frames." By the experimental results by Michelson and Morley, the existence of aether and the appropriateness of Galilean transformation were denied. The experimental results verified that the velocity of light is the same in any inertial frame and supported the relativity principle, which insists that the velocity of light should be the same in all inertial frames.

—◦}}{{◦— ◦}} {{◦ —◦}}{{◦—

Appendix 3.3 Michelson–Morley experiment

As shown in Fig. A3.3, the half mirror HM was set at a 45° angle in the light path from the light source S (Yukawa & Tamura, 1955–1962). The HM split the light, and each new ray proceeded at a mutually 90° angle. Each light ray reflected at M1 and M2, respectively, went back to the HM and then proceeded in the same direction to detector D. Because mirrors M1 and M2 were not precisely perpendicular to the light path, the interference fringe was observed to correspond to the phase contrast of two light rays in a 2D space in the detecting plane. If the measuring system moves at a velocity **v** in the direction of one of two split light rays reflected at the HM, the difference Δt in time in the optical paths of each light ray occurs and the displacement of the interference fringe should be observed to correspond to the difference in time, where the optical path from the HM to the reflective mirror is the same as d. The velocity of light is defined as c, and v/c is β. Then the time difference Δt is given by $d\beta^2/c$. Defining the period of light wave as T, the ratio of displacement of the interference fringe to the interval of neighboring fringes is given by $\Delta t/T$. The wavelength of light is λ. The velocity of light is the distance a light wave travels per second and is given by λv, where v is the frequency of light. Hence the ratio of displacement of the interference fringe to the interval of the neighboring fringes is given by $d\beta^2/\lambda$. In the Michelson–Morley experiment, v was the velocity of the Earth 3×10^4 m/s, d was 11 m, and λ was 5.89×10^{-7} m. Then the ratio of displacement of the interference fringe to the interval of neighboring fringes was estimated to be 0.185 (Michelson & Morley, 1887). Even if the ratio was 1/100 of the estimated one, the ratio should have been measured due to the precision of the Michelson–Morley experiment. But the ratio measured in the experiment was way smaller than the estimated value. In Fig. A3.3 though reflection mirrors M1 and M2 are represented, as mentioned in Section 3.5, in the Michelson–Morley experiment, many reflection mirrors were set to lengthen the optical path by repeated reflection.

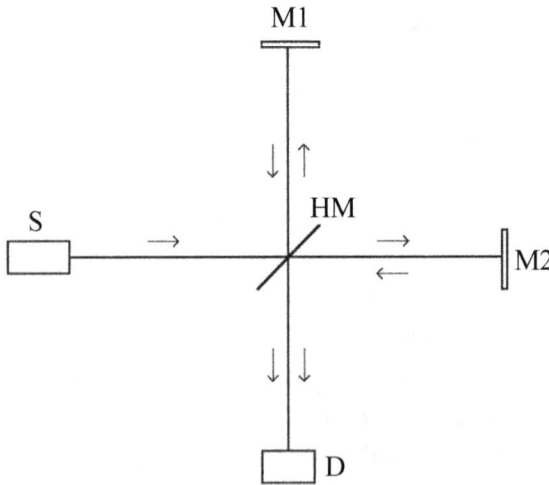

Fig. A3.3 Michelson–Morley experiment. S: light source; D: detector; HM: half mirror; M1 and M2: reflection mirrors.

Fig. 3.13 Albert Abraham Michelson (1852–1931).

Fig. 3.14 Edward William Morley (1838–1923).

3.6 The Special Relativistic Theory

Relativity principle

The historical background of the formulation of the special relativistic theory is given in Appendix 3.4. According to the relativity principle all laws of physics hold true in any inertial frame. In 1905, Einstein published an elementary paper on the special relativistic theory, in which he formulated for the first time a new theory recognizing the relativity principle and derived many results (Section 3.4 in this chapter). According to the relativity principle natural laws appear in the same form in an infinite number of inertial frames mutually moving linearly at a constant velocity. The relativity principle insists on the equivalence of all inertial frames and demands equivalence not only in mechanical phenomena such as Newtonian mechanics but also in electromagnetic phenomena, such as the electromagnetic theory of Maxwell.

Constant light velocity

On the basis of the relativity principle, Maxwell's equations should hold true in the same form in any inertial frame.

—◦》《◦— ◦》 《◦ —◦》《◦—

Appendix 3.4 Background of the formulation of the special relativistic theory

The relativity principle means that all physical laws hold true in any inertial frame. In terms of the elementary laws of electromagnetic theory, the relativity principle was not considered to hold true under Galilean transformation. The reason is as follows: In Maxwell's equations, the velocity of light is supposed to be a constant **c** (equal to the velocity of the electromagnetic wave because light is an electromagnetic wave). Hence by the relativity principle, if Maxwell's equations hold true in any inertial frame, the velocity of light should be a constant **c** irrespective of the light source motion. This contradicts the usual idea of motion. For example, if inertial frame I' moves in the direction of light against inertial frame I, then the velocity of light in I' is smaller than that in I.

Consequently, if the relativity principle is recognized, then the previous idea about transformation between time and space in two inertial frames moving with respect to each other should be improved. But before making any improvement, the necessity of the improvement should be verified. This verification should be obtained only by experimental results, and an optical experiment was the most suitable for this objective.

If the relativity principle is recognized, then aether called as the coordinate system at absolute rest loses physical meaning because when the relativity principle is recognized, the equivalence of all inertial frames is required. This changes the basis of natural description. Einstein improved the previous idea on time-space and formulated a new theory recognizing the relativity principle.

—◦》《◦— ◦》 《◦ —◦》《◦—

Considering these equations involving constant light velocity, light velocity should be constant in vacuum irrespective of whether the light source is moving or not. Then because the velocity of light is constant in any inertial frame, it is invariant for coordinate transformation between inertial frames. Hence it was decided to determine the coordinates of time-space of the inertial frame so that the velocity of light became invariant for coordinate transformation between inertial frames. Galilean transformation could not satisfy this request. Thus Einstein derived the new concept of time-space in an inertial frame (Yukawa & Tamura, 1955–1962).

The conclusions derived from assuming the existence of aether, which was called a "coordinate system at absolute rest," were denied by experimental results. On receiving experimental results verifying the relativity principle, Lorentz thought about what assumption should be introduced so the conclusions estimated assuming the existence of aether suited the relativity principle and for the first time, in 1904, he derived the formula called "Lorentz transformation."

Fig. 3.15 Hendrik Antoon Lorentz (1853–1928).

Einstein independently derived the Lorentz transformation instead of Galilean transformation so that a constant light velocity held true in all inertial frames. Though Lorentz was the first to derive the Lorentz transformation, it was Einstein who derived the transformation based on the fundamental law of the special relativistic theory and indicated the new physical meaning.

3.7 Consequences of the Special Relativistic Theory

Contraction of a moving body

From equations of the Lorentz transformation, the contraction of a moving body was derived (Fig. 3.16). The square root of $[1 - (\mathbf{u}/\mathbf{c})^2]$ was called the Lorentz contraction, where $\mathbf{u}$ is the velocity of the body and $\mathbf{c}$ is the velocity of light. The length of the body in the direction of its movement was given by its length at rest multiplied by the Lorentz contraction (i.e., contracted in the direction of the velocity). When the velocity of the body is way smaller than the velocity of light, the length in the direction of movement is approximately the same as its length at rest.

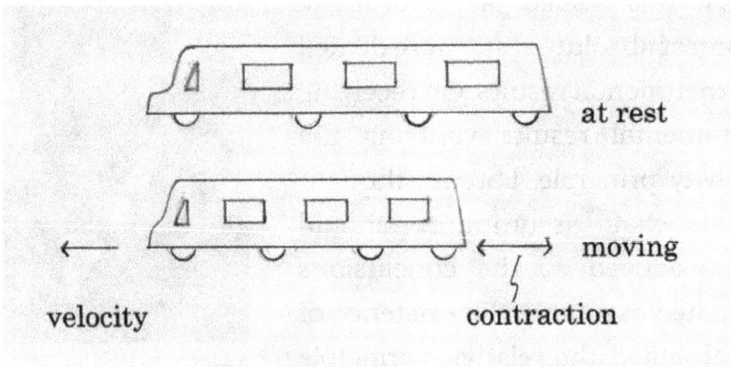

Fig. 3.16 Contraction of a moving body.

Delay of moving clock

Furthermore, from the Lorentz transformation, the delay of a moving clock was derived. The advancement of a moving clock $\Delta\tau$ was given by the advancement of the clock at rest Δt multiplied by the Lorentz contraction (Fig. 3.17). That is, the advancement of a moving clock $\Delta\tau$ is delayed compared to the advancement of a clock at rest Δt. For example, a moving radio-active substance emits radiation for a longer time than the case

time advance

at rest

moving

delay

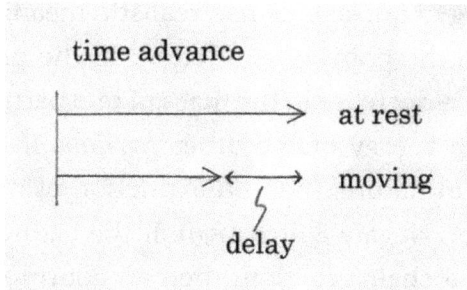

Fig. 3.17 Delay of a moving clock.

at rest because the advancement of a clock is delayed when the substance is moving than when it is at rest and the life time is lengthened.

Relativistic mass of a moving body

Maxwell's equations, which are elemental equations in electromagnetic phenomena, are invariant to the Lorentz transformation and harmonize with the relativity principle. But it was necessary to add a change in the elemental equations of Newtonian mechanics for the purpose of harmonizing the equations of Newtonian mechanics with the relativity principle. From this change, the relativistic mass of a moving body was given by its mass at rest multiplied by the inverse of the Lorentz contraction. It was derived from the law of conservation of momentum.

Mass-energy equivalence

From the relation in the Lorentz transformation between momentum $\mathbf{p}$ and energy E, Einstein indicated that the relativistic mass was given by $E/\mathbf{c}^2$ and indicated the mass-energy equivalence. Hence, according to Einstein, a mass point with mass m has energy E, of

$$E = m\mathbf{c}^2 .$$

The mass-energy equivalence has realistic meaning only when a particle is on the point of extinction. In the process, energy equivalent to the decrease in the mass of one particle is changed into the kinetic energy of the other particle. If the extinction process exists in nature, then on extinction of mass, energy is released (Møller, Nagata & Ito, 1959). If the extinction of a particle occurs in a chain reaction, then an enormous amount of energy is released. After hearing horrific tales of Nazi atrocities, in 1939, Einstein suggested to President Franklin D. Roosevelt the possibility of the production of an atomic bomb on the basis of this theoretical conclusion.

Mass defect

Mass-energy equivalence was experimentally verified in a nuclear fission reaction with a different mass defect (Explanation 3.1). Mass-energy equivalence was verified in an experiment by John Douglas Cockcroft (Fig. 3.19) and Sinton Walton (Fig. 3.20) at the Cavendish Institute. In the experiment, when a proton hit a lithium atom with a high velocity, the proton plunged into the lithium's nucleus and a compound nucleus was formed. The new nucleus formed by the addition of an incident particle to the nucleus, whose interior energy is higher with added energy, is called a compound nucleus (Fig. 3.18).

In the experiment, because the compound nucleus was unstable, it divided into two high-velocity a particles (helium

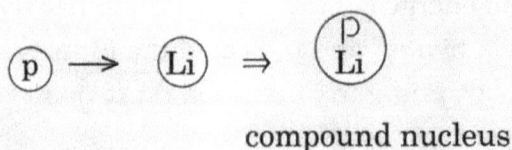

compound nucleus

Fig. 3.18 Compound nucleus. p: proton; Li: Lithium.

nucleus). By measuring the lost mass in the nuclear fission reaction and the quantity that was obtained by subtracting the incident proton's kinetic energy from the α particle's kinetic energy obtained after the nuclear fission reaction, it was found that the energy equivalent to lost mass coincided with kinetic energy. Thus, mass-energy equivalence was verified.

—»⟩⟨⟨·— ·⟩⟩ ⟨⟨· —·⟩⟩⟨⟨·—

Explanation 3.1 Mass defect

The difference between the real mass of a nucleus and the mass of the sum of the protons and neutrons that constitute the nucleus is called the mass defect (Fig. E3.1). The energy necessary to break up a nucleus into its constituent particles is called its binding energy.

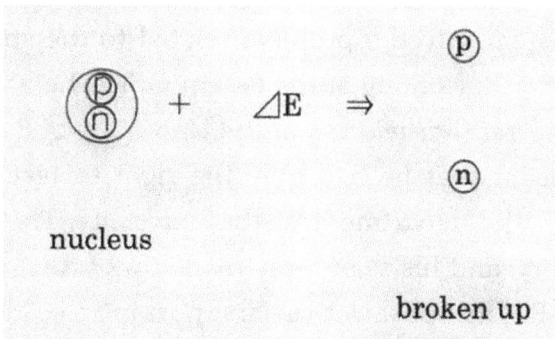

Fig. E3.1 Binding energy. ΔE: binding energy; p: proton; n: neutron.
The nucleus is broken up by ΔE.

When the binding energy is expressed by ΔE, the mass defect is given by $\Delta E/c^2$. The sum of the mass of the particles making up the nucleus after they break up is not equal to the original nucleus's mass, and the difference in masses corresponds to the binding energy. Einstein indicated that the difference in mass is given by the ratio of binding energy to the square of light velocity from mass-energy equivalence (Møller, Nagata & Ito, 1959).

—»⟩⟨⟨·— ·⟩⟩ ⟨⟨· —·⟩⟩⟨⟨·—

Fig. 3.19 John Douglas Cockcroft (1897–1967).

Fig. 3.20 Ernest Thomas Sinton Walton (1903–1995).

3.8 Research at the University

Getting a post at the university

In 1907, Einstein solved a problem related to the specific heat of solids. In 1908, working at the Patent Office, he got a post as a private lecturer (*privatdozent*) at the University of Bern. Taking the post meant that he had only the right to teach, without belonging to the department of the university. The university paid no salary, and his salary was made up of the fees paid by attendants. But this post was the first post in an academic place for him. Then his sister, studying at the University of Bern, attended his lecture. On December 21, 1908, she submitted a thesis on romance languages to the University of Bern and got a PhD degree in romance languages magna cum laude.

On October 15, 1909, he became an associate professor of theoretical physics at the University of Zurich. He resigned from the University of Bern and the Patent Office. It was then he was recognized as a leading scientist. He was invited to be a professor at the Karl-Ferdinand University in Prague. In March 1911, he and his family arrived at Prague and he joined on April 1.

In 1907, Grossmann became a full professor of geometry at ETH. In 1911, he became dean of the mathematics-physics section of ETH. The young dean's first work was to sound Einstein out on his thought on returning to Zurich, this time to the ETH (Pais, 1982, pp. 208–209). Einstein immediately informed him of his thought to teach at ETH.

In 1912, he was invited as professor at ETH and he relocated to Zurich from Prague. Before he relocated to Zurich, the University of Utrecht invited him, but he rejected the offer. Though on graduating ETH in 1900, he had not succeeded in getting a post at the university, 12 years after graduation, he joined as professor at ETH.

Founding the general relativistic theory with Grossmann

1912 onward, with Grossmann, Einstein set about founding the general relativistic theory, which generalized the special relativistic theory. Using tensor calculus of Tullio Levi-Civita and Gregorio Ricci-Curbastro and Riemannian geometry, which was theory about curved space and surface, he developed gravity theory. In 1913, he published a paper on the general relativistic theory entitled "Draft of the General Relativistic Theory and Gravity Theory" [*Entwurf Einer Verallgemeinerten Relativitätstheorie und Einer Theorie der Gravitation*] with Grossmann. In the paper, they proposed an idea that gravity occurred due to distortion of time-space. At the time, they had not come up with the gravitational field equation that related gravity to the distortion of time-space. Einstein came up with the equation in 1915.

In April 1914, because the University of Berlin invited Einstein to be a professor after nomination by Planck, he relocated to Berlin from Zurich. Immediately afterward, his wife and two sons (Hans Albert and Eduard) came to him. But they went back to Zurich, and he started living apart from his family, devoting himself to theoretical research. For 18 years, he stayed in Berlin. He joined the Kaiser Wilhelm Institute for physics as director. Though he came back to Germany, he did not intend to get citizenship of Germany in the situation, already having the citizenship of Switzerland in 1901. He wrote in a letter to Lorentz that "A post in Berlin frees me of all obligations so that I can devote myself freely to thinking." He wrote in a letter to his friend Heinrich Zannger, director of the Institute for Forensic Medicine at the University of Zurich, that "Contact with the colleagues in Berlin might be stimulating, and especially the astronomers are important for me," indicating that Einstein then was interested in the bending of light (Pais, 1982, p. 240).

On July 2, 1914, at the Prussian Academy, he delivered an inauguration lecture as professor at the University of Berlin. He admired Planck as Planck's idea of an energy quantum had introduced improvements in the microscopic field and then delivered a lecture on his own relativity theory. In reply to his lecture, Planck finished his speech by saying that experimental results on the bending of light due to gravity theoretically predicted by Einstein are expected to be obtained by the total solar eclipse on August 21 (Pais, 1982, p. 242). However, the First World War broke out August 1, preventing people from checking the claim. Though the prediction of the bending of light due to gravity was not correctly explained at the time, the prediction was made correctly in the complete version of the general relativistic theory in 1915, and the correctness of the prediction was verified by observing the total solar eclipse in 1919 (Section 3.11).

As mentioned above, on August 1, 1914, the First World War commenced. Einstein objected to war. He appealed to people, saying that it was stupid to slaughter each other for the purpose of the nation. However, none listened to him and young people were sent to the battlefield. The First World War, the first of its kind in human history, resulted in widespread death and destruction.

David Hilbert

On November 25, 1915, Einstein submitted the complete version of the general relativistic theory to *Annalen der Physik*. Five days before his submission, David Hilbert (Fig. 3.21) submitted a paper that contained gravitational field equations of the general relativistic theory, and there could have been questions regarding who was the first one to come up with the equations. The fact is, when visiting Gottingen, Einstein had discussed his idea with Hilbert. Afterward Hilbert came up with the gravitational field equations several days before Einstein did. Nevertheless, Einstein's got rec-

Fig. 3.21 David Hilbert (1862–1943).

ognized for the equations because he related gravity to the distortion of time-space (Hawking & Sato, 2001, p. 30). In 1916, the submitted paper "Basis of the General Relativistic Theory" [*Die Grundlage der Allgemeinen Relativitätstheorie*], the first and most important paper on the subject, was published.

3.9 The General Relativistic Theory

General relativity principle

The special relativity principle means equivalence of all inertial frames moving at a mutually constant velocity and without distortion. In other words, it means that the fundamental equations in physics hold true in the same form in any inertial frame. On the other hand, the general relativity principle means the equivalence of not only the inertial frames but also the accelerative frames (noninertial frames). Accelerative frames do not move at a constant velocity but with acceleration against inertial frames. For example, a system rotating against inertial frames and with a centrifugal force is an accelerative frame. On the basis of the general relativity principle, Einstein formulated the general relativistic theory generalizing the special relativistic theory.

In place without mass and without gravity, when an accelerative frame rotates against inertial frames, gravity occurs in the appearance of the centrifugal force. Einstein introduced the equivalence principle, which insists equivalence between "gravity in appearance" and gravity.

Minkowski's 4D world

To describe natural phenomena, Einstein used Minkowski's 4D world, where 3D space coordinates (X, Y, Z) and the time coordinate (T) in the inertial frames without distortion were unified to X_i (i = 1, 2, 3, 4) = ($X, Y, Z, \mathbf{c}T$, where $\mathbf{c}$ is the velocity of light). Minkowski's 4D world was called the "World" (Explanation 3.2). The square of a segment (4D distance between two neighboring points) in Minkowski's 4D world without distortion was given by the subtraction of the square of segment dX_4 (= $\mathbf{c}dT$) on the time coordinate X_4 (= $\mathbf{c}T$) from the sum of three squares of segment dX_i (i = 1, 2, 3) (1D distance between two neighboring points). On the other hand, in accelerative frames, the segment in Minkowski's 4D world expressed by the general coordinate x_i (i = 1, 2, 3, 4) with distortion was given by the product of two segments dx_i (i = 1, 2, 3, 4) of the general curve coordinate multiplied by the coefficient g_{ij}, that is, $g_{ij}dx_idx_j$ (i, j = 1, 2, 3, 4), which was total summation for all pairs (i, j). The coefficient g_{ij} (i, j = 1, 2, 3, 4) was called the metric tensor. The general curve coordinate x_i (i = 1, 2, 3, 4) is a function of X_i (i = 1, 2, 3, 4). In Minkowski's 4D world without distortion, it holds true that

$x_i = X_i$ (i = 1, 2, 3, 4), $g_{11} = g_{22} = g_{33} = 1$, $g_{44} = -1$, and g_{ij} with different i and j is zero. The metric tensor plays an important role in the general relativistic theory, as mentioned below.

— ❧❦ — ❧ ❦ — ❧❦ —

Explanation 3.2 Minkowski's 4D world

The fusion of spatial coordinates (X, Y, Z) and the time coordinate T is called Minkowski's 4D world. Here, it is assumed that there is no distortion of time-space. Four coordinate axes are mutually perpendicular. In Fig. E3.2, for convenience, 3D spatial coordinate axes (X, Y, Z) are expressed by a plane perpendicular to the time axis T. The trajectory of a body at rest is parallel to the time coordinate axis in Minkowski's 4D world. The trajectory of an event traveling in a future direction at the velocity of light is the upper cone in Fig. E3.2. The trajectory of an event traveling from the past is the lower cone. This cone is called the light cone.

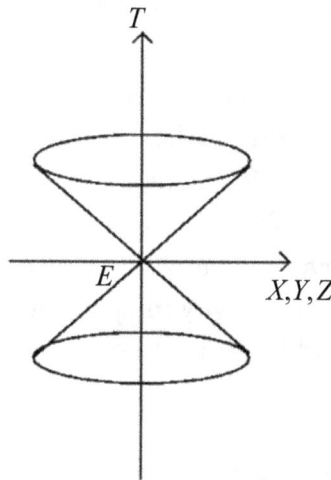

Fig. E3.2 A light cone.

Because the velocity of a body is less than the velocity of light, the world line, which is the trajectory of a moving body in Minkowski's 4D world, exists inside the light cone.

— ❧❦ — ❧ ❦ — ❧❦ —

The gravitational field equation that decides the variable (gravitational potential) of the gravitational field and the metric tensor was created by Einstein in 1915, as mentioned in this section. The gravitational field equation contains terms concerning the distortion of time-space due to gravity and the term of mass-energy. In the case of a weak gravity field and static mass distribution, the gravitational field equation becomes Newton's gravitational theory, which is expressed by the term of gradient of gravitational potential and the term of mass distribution.

The metric tensor determines the geometry (generally, non-Euclidean geometry) of Minkowski's 4D world in an accelerative frame. It was indicated that the metric tensor depended on gravity in appearance (like centrifugal force) and gravity. In other words, gravity influences metric tensor and causes distortion of time-space.

In the world where Minkowski's 4D world is not flat and time-space distorts due to gravity, the path giving the minimum distance between two points is not a linear line but a curve. In the real world, Euclid geometry does not hold true and the sum of the interior angles of a triangle is smaller than 180 degrees (π radian). When an accelerative frame with gravity in appearance, such as centrifugal force, transfers to an inertial frame with no rotation, gravity in appearance extinguishes and Minkowski's 4D world becomes flat.

World line

A point of Minkowski's 4D world, that is, a point representing "when and where" is called the "world point." The moving path of a particle in Minkowski's 4D world is called the "world line." The world line where the distance between two world points in Minkowski's 4D world is the minimum possible value is called the "geodesic line." To generalize the linear line giving the minimum distance line in the world where there is no distortion of time-space as the inertial frame and Euclid geometry holds true, a geodesic line is defined as the minimum distance line in the general relativistic theory. It was indicated that the world line of a particle in a free fall only under gravity was the geodesic line. The geodesic line was given as a solution to the Euler equation derived from the variational principle. Though in Newtonian mechanics, the law of inertia is the law of motion of a body without force, in the general relativistic theory, the law of inertia is replaced by "a body only under gravity moves along geodesic line."

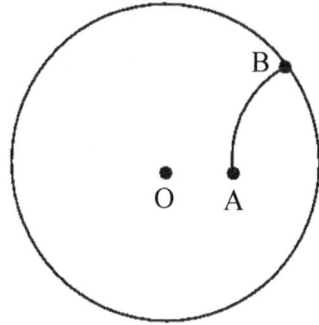

Fig. 3.22 A geodesic line in a rotating accelerative frame. A geodesic line, which is the shortest path connecting points A and B, is not linear but a curve in an accelerative frame rotating around point O.

Gravitational acceleration is derived from gravitational potential. As gravitational potential, there are gravitational scalar potential and gravitational vector potential (3D vector defined using the metric tensor). When the spatial coordinate system is perpendicular to the time coordinate axis, or the gravitational vector potential does not depend on time, the gravitational acceleration is given by the spatial gradient of the gravitational scalar potential multiplied by -1. For example, in accelerative frames rotating with a rotation angular velocity ω, the gravitational scalar potential of a particle located at radius r is given by $-(r\omega)^2/2$. Hence, gravity acceleration is $r\omega^2$ and its direction is the direction of increasing r, that is, it is equal to the acceleration of the centrifugal force. A metric tensor appearing in the segment of a geodesic line is influenced by gravity and distortion of time-space occurs, and a geodesic line is not a linear line but a curve, as shown Fig. 3.22.

3.10 Consequence of the General Relativistic Theory

Relation between metric tensor and gravity

An element g_{44} of the metric tensor was expressed by $-(1 + 2\chi/\mathbf{c}^2)$, where χ is the gravitational scalar potential. Though χ is expressed by (χ + constant) giving the same gravitational acceleration, χ is normalized so that the constant is determined for g_{44} to be -1, which is the value of no gravitation. Thus, the general relativistic theory related a metric tensor in Minkowski's 4D world to gravity. Because a metric tensor depends on gravitational scalar potential in the general curve coordinate with gravity, the metric tensor changes with time or space. Hence, distortion, which is defined as the differential calculus of the metric tensor with respect to time and space, occurs. When the angular velocity of a rotating accelerative frame changes to zero and transfers to the inertial frame, the gravitational scalar potential becomes zero and the metric tensor becomes constant, as mentioned above. Hence, the distortion of time-space extinguishes and the world becomes flat.

Einstein insisted on the equivalence of a permanent gravitational field produced by a large mass, such as the Earth or the Sun, and a nonpermanent gravitational field produced artificially, such as centrifugal force, by the equivalence principle.

Relativistic mass

In the general curve coordinate with gravity, the mass of a particle was expressed by relativistic mass. This depended on the gravitational scalar potential and the gravitational vector potential. If the space coordinate system is perpendicular to the time coordinate axis, the gravitational vector potential is zero and the relativistic mass is expressed by the product of mass at rest by the inverse of the Lorentz contraction $\{1 + (2\chi - \mathbf{u}^2)/\mathbf{c}^2\}^{1/2}$, which is given by generalizing the Lorentz contraction in the special relativistic theory by considering χ, where $(\cdot)^{1/2}$ represents the square root and $\mathbf{u}$ is the velocity.

Total energy

The Euler equation, which determines the world line of a particle in Minkowski's 4D world, consists of four equations. The first three are motion equations, represented in a form depending on the metric tensor. The fourth equation expresses the energy conservation law. From the fourth equation, in the case of a weak gravitational field, neglecting the squared terms of $\mathbf{u}/\mathbf{c}$ above and leaving till the square of $\mathbf{u}/\mathbf{c}$, the total energy of a particle in a stationary gravitational field is given as follows:

$$H = m_0 \mathbf{c}^2 + m_0 \mathbf{u}^2/2 + m_0\chi \,.$$

The first term represents static energy, the second term represents kinetic energy, and the third term represents gravitational potential energy. When $\mathbf{u}$ is large, the energy cannot be separated into kinetic energy and potential energy. The first one is a special term in relativity theory due to the equivalence of mass and energy.

Advancement of a moving clock

The advancement $\Delta\tau$ of a clock located at a particle moving at a velocity $\mathbf{u}$ in a gravitational field expressed by the gravitational potential was given by the product of advancement Δt of the clock at rest in inertial frames and the generalized Lorentz contraction. From this, the clock moving in a gravitational field is delayed (or advances). This is the generalization of delay in the special relativistic theory and depends on the velocity $\mathbf{u}$ and χ. When the clock is at rest in the frames under a gravitational field, the velocity $\mathbf{u}$ is zero and $\Delta\tau$ depends only on χ, and the Lorentz contraction is $(1 + 2\chi/\mathbf{c}^2)^{1/2}$. For example, when the clock is at rest at radius r on a disk rotating with an angular velocity ω, χ is $-(r\omega)^2/2$, and the Lorentz contraction is $\{1 - (r\omega)^2/c^2\}^{1/2}$, the clock is delayed far from the center. An observer standing on the disk will think that the delay of the clock is caused by the existence of a gravitational scalar potential due to a centrifugal force in the disk. On the other hand, an observer standing on the inertial frames will think that the delay of the clock is caused by the velocity **u** of the clock (velocity **u** is equal to $r\omega$) because there is no gravitational field and χ is zero.

3.11 Verification of Correctness of the General Relativistic Theory

Bending of light

In a stationary gravitational field, light velocity depends on gravitational potential. Because the path of light complies with Fermat's principle (the path is determined so as to minimize the total time necessary for passing through the path), the light path is different from a linear line due to gravity, according to the equation of light path derived from Fermat's principle. That is, Einstein concluded that there exists the "phenomenon of the bending of light" (Explanation 3.3). He calculated the angle of the bending of light passing near the Sun and predicted that the angle of bending was 1.7″. This prediction could be verified by observing the total solar eclipse without the influence of sunlight. Arthur Stanley Eddington's (Fig. 3.24) observation results are described below.

Perihelion motion of Mercury

A phenomenon other than the "bending of light due to gravity" used to verify the correctness of the general relativistic theory was the "perihelion motion of Mercury." This phenomenon could not be explained by Newtonian mechanics. Considering a planet moving in the gravitational field of the Sun being much heavier than Mercury, Einstein estimated that the perihelion of Mercury moves due to the gravitational field of the Sun and predicted that the angular change per 100 years was 43", representing the location of Mercury viewed by the Sun by the angle in the revolution plane. This prediction coincided with the observation results.

— ·⟩⟩ ⟨⟨· — ·⟩⟩ ⟨⟨· — ·⟩⟩ ⟨⟨· —

Explanation 3.3 Path of light in a gravitational field

In a coordinate system where the time axis is perpendicular to the spatial coordinate system, we consider the propagation of light in a static gravitational field (Møller, Nagata & Ito, 1959). Then, the dielectric constant, the magnetic permeability, and the refractive index given by the square root of the product of the dielectric constant and magnetic permeability depend on the gravitational scalar potential. Hence, the propagation velocity of light w that is given by the velocity of light divided by the refractive index depends on the gravitational scalar potential.

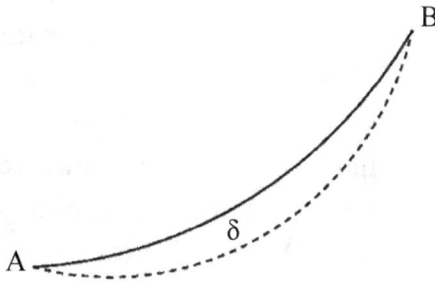

Fig. E3.3 Path of light passing from A to B in a gravitational field (the broken line is the new path with variation δ).

According to Fermat's principle, the path of light passing from A to B, as shown in Fig. E3.3, is the path that requires the minimum time to navigate. This path is given by the light path decided by the solution of the Euler equation derived from the variational principle minimizing the sum of time $d\sigma/\mathbf{w}$ for the propagating spatial segment $d\sigma$. Because the Euler equation contains the gradient of the gravitational scalar potential, the light path depends on gravitation, that is, though the light path is a linear line when there is no gravitation, when there is gravitation, the light path is not a linear line but results in the "phenomenon of bending of light due to gravitation." Einstein expressed the bending angle of light passing near the Sun as $4kM/(c^2 r_m)$, where k is the universal gravitational constant, M is the mass of the Sun, c is the velocity of light, and r_m is the radius of the Sun. From this, he predicted that the bending angle is 1.7" (Møller, Nagata & Ito, 1959).

—◦◊◊◦•— ◦◊◊ ◊◊◦ —◦◊◊◦•—

Spectrum shifts

Because the advancement of a clock depends on the gravitational potential, Einstein estimated that if the difference between gravitational scalar potentials at the light emitting place and the light observing place is represented as $\Delta\chi$, then the spectrum is displaced depending on $\Delta\chi$, that is, when the frequency of the light observed is v, with the proper frequency of an atom as v_0 and Δv is defined as the difference of frequencies $v - v_0$, then the ratio of $\Delta v/v_0$ is given by $\Delta\chi/c^2$. He predicted that in the case of the gravitational field of the Sun, $\Delta v/v_0$ is -2.12×10^{-6}. When the light emitted from an atom on the surface of the Sun is observed on Earth, the spectrum of light has a slightly smaller frequency than the spectrum of light emitted from an atom on the Earth, that is, the spectrum shifts to red. Einstein's prediction on the spectrum's shift sufficiently coincided with the observed results of the Sun and Sirius Binary star.

On the general relativistic theory, there is little experimental verification. Einstein's general relativistic theory contains Newtonian mechanics as the first approximation. The fact that the difference between Newtonian theory and Einstein's theory appears in only three phenomena (shift of spectrum due to gravitation, bending of light due to gravitation, and perihelion motion of Mercury) indicates that Newtonian theory is a good approximation of gravitational phenomena in the solar system. But on treating cosmic phenomena, it is expected that Einstein's general relativistic theory is a leading guide. For example, the theory of black holes, which is the final stage of a star's evolution, was developed on the basis of the Schwarzschild solution of the gravitational field equation.

The total solar eclipse observed by Eddington

Because the general relativistic theory could explain the perihelion motion of Mercury, which was considered a mystery because of the impossibility of the theoretical explanation, Eddington was very interested in this theory. He was an enthusiastic commentator that he delivered a lecture on the general relativistic theory at a meeting of the Royal Society in 1916. In order to verify the phenomenon of bending of light due to the Sun's gravitation (Fig. 3.23), as predicted by Einstein, in 1919, he led members of the total solar eclipse observation group

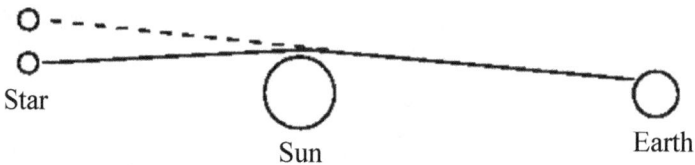

Fig. 3.23 Bending of light due to gravitation. During a total solar eclipse, a star whose light passes near the Sun appears to be in the direction of the broken line for an observer on the Earth.

Fig. 3.24 Arthur Stanley Eddington (1882–1944).

to Principe island. On November 6, at the joint meeting of the Royal Society and the Royal Astronomical Society, from analysis of observation results, it was shown that the bending angle of light passing near the Sun was 1.61" ± 0.30", verifying the correctness of the value 1.7" predicted by Einstein. On November 7, carrying the headline "Revolution in Science: New Theory of Cosmos," *The Times* introduced Einstein's general relativistic theory and reported Eddington's observation results indicating the correctness of the theory. This created a great sensation, and Einstein came to be known across the world.

—•》《•— •》 《• —•》《•—

Explanation 3.4 Analysis of the perihelion motion of Mercury

We consider the motion of a planet as a mass point in a weak gravitational field. We assume that mass distribution producing a gravitational field is static (meaning independent of time) and spherically symmetric. The Sun corresponds to it. We use polar coordinates (r, θ, φ), as shown in Fig. E3.4, where the original point is the place of the Sun.

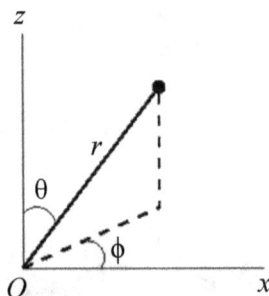

Fig. E3.4 Polar coordinates (r, θ, φ).

We consider the motion of a planet with mass m in the gravitational field of the Sun, with mass M. From the solution of Newton's theory, the planet moves in an elliptic orbit where the distance between the planet and the Sun is r, θ is $\pi/2$ (which means the planet is in the x-y plane), the eccentricity is ε, the aphelion r_1 is $d/(1 - \varepsilon)$, the perihelion r_2 is $d/(1 + \varepsilon)$, and d is given by b^2/a, where the semimajor axis a is $(r_1 + r_2)/2$ and the semiminor axis b is $(r_1 \times r_2)^{1/2}$.

The eccentricity ε is given by f/a, where f is the distance between the center of the ellipse and the focus. In the case of Mercury, $\varepsilon = 0.2056$ and $d = 5.786 \times 10^{10}$.

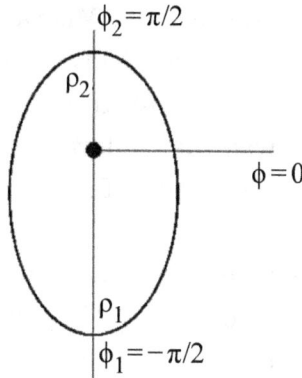

Fig. E3.5 Aphelion ρ_1 and perihelion ρ_2.

As shown in Fig. E3.5, when ρ_1 and ρ_2 are defined as $1/r_1$ and $1/r_2$, respectively, the increase of φ from the aphelion to the perihelion is given as follows because $\varphi_1 = -\pi/2$ at the aphelion ρ_1 and $\varphi_2 = \pi/2$ at the perihelion ρ_2.

$$\varphi_2 - \varphi_1 = \pi.$$

On the other hand, on the basis of $(\varphi_2 - \varphi_1)$ obtained from the equation determining the orbit of a planet strictly according to Einstein's theory, we consider the difference $2(\varphi_2 - \varphi_1)$ of φ between two continuous perihelions. The difference $\Delta\varphi$ between the result when strictly using Einstein's theory and the result of $2(\varphi_2 - \varphi_1) = 2\pi$ in Newton's theory was given by $3\pi a(\rho_1 + \rho_2)/2$, where a is defined as $2kM/c^2$, with k being the universal gravitational constant and c being the velocity of light.

The result when strictly using Einstein's theory is larger with $\Delta\varphi$ than the result when using Newton's theory approximately, and a positive $\Delta\varphi$ means that with every revolution of a planet, the perihelion advances. In the case of Mercury, the advance of the perihelion predicted by Einstein was by an angle of 43" per 100 years (Møller, Nagata & Ito, 1959, p. 348). The value coincided with the observation result. In the case of other planets, the advance of the perihelion is too small to identify with any certainty.

—◊≫ ≪◊— ◊≫ ≪◊ —◊≫ ≪◊—

3.12 The Nobel Prize

Elsa

Immediately after leaving his wife Mileva, Einstein formulated the gravitational field equation (Section 3.9). After five years of separation, on February 14, 1919, the couple divorced. A condition of the divorce was that Mileva would receive the prize money of the Nobel Prize that Einstein would be awarded (Pais, 1982, p. 300).

In 1917, Einstein married Elsa Lowenthal (Figs. 3.25 and 3.26), who was a cousin and a divorcee. They had known each other since childhood. She took care of him when he fell seriously ill in 1917. She was born in 1876. Her parents lived on the lower floors in the same building. Her father was the first cousin of Hermann, Albert's father, and her mother was a sister of Pauline, Albert's mother. Elsa—gentle, warm, motherly, and prototypically bourgeois—loved to take care of her Albert. She fell in love with Albert because he played Mozart so beautifully on the violin. Einstein wanted the right to have his sons from

Fig. 3.25 Einstein and Elsa.
(Photograph taken in 1921.)

Mileva visit him in Berlin. He sent a note of thanks to Mileva saying, "I am likewise thankful that you have not alienated me from the children" (Isaacson, 2017, pp. 217, 228, 229).

In January 1920, Einstein's mother came to Berlin, wanting to live with her son. But she was ill with abdominal cancer and passed away in February. In 1918, the First World War ended. The victorious allied powers imposed severe con-

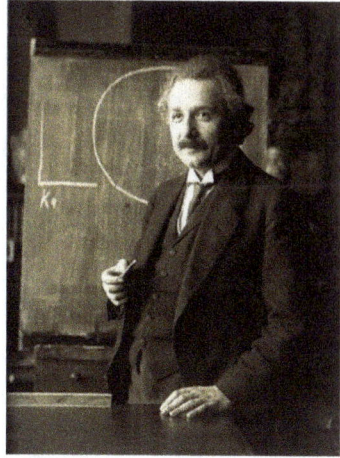

Fig. 3.26 Einstein in 1921.

ditions on a defeated Germany, pushing the country into extreme economic distress. The year Einstein lost his mother, seeing Germany's downfall, he—a German by birth—again got the citizenship of Germany, which he had given up at the age of 16, in 1920.

Visiting the United States

In 1921, Einstein visited the United States for the first time. The main object was to raise funds to create a Hebrew university in Jerusalem. So, a noisy parade was taken out. The route was lined with more than 15,000 spectators, and the crowds cheered wildly (Isaacson, 2017, p. 300). It was rather unusual for a theoretical physicist. They hoped to raise at least $4 million. By the end of the year, only $750,000 had actually been collected (Isaacson, 2017, p. 300). During his stay in the United States, Einstein was requested to deliver lectures. During a lecture Einstein delivered at Princeton, he remarked, "Nature hides her secret because of her essential loftiness, but not by means of ruse." Mathematician Oswald Veblen requested Einstein that they be permitted to carve the words on the stone mantel of the fireplace in the common room in a new mathematics building being planned to be completed a decade later. Einstein sent back his approval. Afterward, the building became the home of the Institute for Advanced Study. In 1933, Einstein decided to immigrate to Princeton and had an office there. He sat in front of the fireplace in the latter part of his life (Isaacson, 2017, pp. 297–300).

Travel abroad for safety

On June 24, 1922, Einstein's friend Walter Rathenau, who had served as the foreign minister of Germany for a few months, was assassinated by the ultraright organization Consul. On July 4, sensing danger to his life, Einstein wrote a letter to Marie Curie, saying that he should resign from his post in Preussen Science Academy (Pais, 1982, p. 316). On October 8, he traveled abroad for safety. On route, Einstein received word that he had been awarded the Nobel Prize for his research on the photoelectric effect. He was awarded the prize in the same year.

The prize money was given to Mileva, his first wife, according to the divorce condition. In February 1923, several months after his departure, Einstein came back to Berlin. During his travel, he stayed for short periods in Colombo, Singapore, Hong Kong, and Shanghai. He stayed in Japan for five weeks and in Palestine for 12 days. Every place he went, enthusiastic crowds

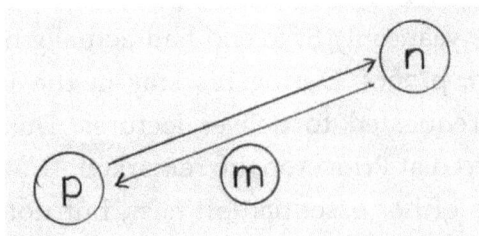

Fig. 3.27 A meson. p: proton; n: neutron; m: meson (carrier of the nuclear force).

welcomed him. When Hideki Yukawa (Fig. 3.28) was the 4th student of Kyoto, the first junior high school of the former system, one year before entering the third high school of the former system, Einstein visited Japan. There was massive excitement among the people. In his autobiography *Tabibito* [*The Traveler*], Yukawa wrote that a friend on coming to know of Einstein's visit, while doing a physical experiment with a partner of

Fig. 3.28 Hideki Yukawa (1907–1981). (Photograph taken in 1949.)

Yukawa, said, "Ogawa (former name of Yukawa) would become a person like Einstein" (Yukawa, 2011, p. 143). Afterward, Yukawa proceeded to Kyoto University and in 1935 published the paper "Theory of Meson" (Fig. 3.27), which played an important role in interpreting the mechanism inside of a nucleus, and predicted the existence of the meson, a carrier of nuclear force (the force between a proton and a neutron) (Zee, 2010, p. 28).

The research result was accomplished through self-supported ardor in Japan, where there was no leader in the field, and was superior to the first class of research in Europe (Nambu, 1998, p. 71). In 1949, Yukawa was awarded the Nobel Prize in physics for research for the first time in Japan. Yukawa would meet with Einstein later in his life at the Institute for Advanced Study (Section 3.14).

In 1925, Einstein was awarded the Copley Medal by the Royal Society, and in 1926, he was awarded the Gold Medal of the Royal Astronomical Society.

Einstein and Germany

Fame attracts envy and hatred. Einstein was no exception. In this instance, these hostile responses were particularly intensified because of his exposed position in a turbulent environment. During the 1920s, he was a highly visible personality, not for one but for a multitude of reasons.

On May 5, 1916, he succeeded Planck as the president of Deutsche Physikalische Gesellschaft. In 1917, he began his duties as the director of the Kaiser Wilhelm Institut für Physik. In 1922, the academy appointed him to the board of directors of the astrophysical laboratory in Potsdam. On February 12, 1920, disturbances broke out in the course of a lecture given by Einstein at the University of Berlin.

On August 24, 1920, a newly founded organization, the Arbeitsgemeinschaft Deutscher Naturforscher, organized a meeting in Berlin's largest concert hall for the purpose of criticizing the content of relativity theory and the alleged tasteless propaganda in its favor by its author (Pais, 1982, pp. 312–316).

3.13 The Fifth Solvay Conference

In October 1927, the Fifth Solvay Conference was held (Fig. 3.35) and there was a discussion on quantum mechanics, which was one of the two great theories in modern physics along with the relativity theory. Famous scientists working on quantum mechanics attended the conference, including Planck, Niels Hendrik David Bohr (Fig. 3.29), Louis-Victor Pierre Raymond Duc de Broglie (Fig. 3.30), Werner Karl Heisenberg (Fig. 3.31), Erwin Rudolf Josef Alexander Schrodinger (Fig. A3.4), Paul Adrian Maurice Dirac (Fig. 3.34), and Einstein.

Survey on works of scientists developing quantum mechanics

To survey works of scientists developing quantum mechanics, it will be useful to understand the foundation of quantum mechanics. First, Planck proposed the concept of energy quantum and played a leading role in quantum mechanics, as mentioned before. Einstein verified the correctness of Planck's concept by theoretical elucidation of the photoelectric effect (Section 3.4).

In 1913, Bohr proposed the model of the structure of an atom. In 1924, de Broglie insisted that all particles have the property of a wave and gave the wavelength by dividing Planck's constant by the momentum of a particle. He was also successful in giving a physical explanation of Bohr's atomic structure model.

Fig. 3.29 Niels Hendrik David Bohr (1885–1962). (Photograph taken in 1922).

Fig. 3.30 Louis-Victor Pierre Raymond Duc de Broglie (1892–1987). (Photograph taken in 1929).

Fig. 3.31 Werner Karl Heisenberg (1901–1976). (Photograph taken in 1927.)

Fig. 3.32 Wolfgang Ernst Pauli (1900–1958). (Photograph taken in 1922.)

Fig. 3.33 Max Born (1882–1970).

Fig. 3.34 Paul Adrian Maurice Dirac (1902–1984). (Photograph taken in 1933.)

Fig. 3.35 Photograph at the Fifth Solvay Conference.

In 1926, influenced by de Broglie proposal of the wavelike property of a particle, Schrodinger (Appendix 3.5) derived the Schrodinger equation satisfied by the wave function representing the wavelike property of an electron and formulated the wave mechanics. In 1925, one year earlier, Heisenberg had formulated matrix mechanics independently of Schrodinger. Wave mechanics led to the formulation of quantum mechanics with a wave function. On the other hand, matrix mechanics led to the formulation of quantum mechanics with a matrix representation. Both theories were different in their methods of expression but were equivalent. Thus, both theories contributed to the foundation of quantum mechanics.

Wolfgang Ernst Pauli (Fig. 3.32) founded the exclusion principle. In quantum mechanics, the state of an electron is represented by its discrete state, determined by the following: (i) energy, (ii) orbital angular momentum (corresponds to the electron's revolution around the nucleus), and (iii) spin (corresponds to the electron's rotation). The exclusion principle meant that there could be only one electron in one quantum state.

A particle following the exclusion principle, such as an electron, is called a fermion and a particle not following the principle, such as a photon, is called a boson.

In 1926, Max Born (Fig. 3.33) discovered that the square of the absolute value of the wave function, which is the solution of the Schrodinger equation, meant the probability of the presence of a particle, giving physical meaning of wave function for the first time. In 1928, Dirac formulated the theory unifying the relativity theory and quantum mechanics. He derived the Dirac equation concerning spin.

Probabilistic interpretation in quantum mechanics

During the conference, all attendees stayed in the same hotel and discussions happened during meals. This was when a major debate broke out between Bohr and Einstein. On one hand, Einstein insisted that probabilistic interpretation in quantum mechanics is uncertain, like the roll of a dice, and that God does not play dice with the universe (Hawking & Sato, 2001, p. 37).

On the other hand, Bohr considered that probabilistic interpretation in quantum mechanics is appropriate. The great debate between Bohr and Einstein continued for four years. In February 1931, Einstein finally accepted Bohr's criticism and his thought on quantum mechanics changed radically (Pais, 1982, p. 448). Consequently, in September of the same year, he sent a letter to the Nobel committee in which he nominated Heisenberg and Schrodinger for the Nobel Prize.

—◦⟩⟨◦— ◦⟩ ⟨◦ —◦⟩⟨◦—

Appendix 3.5 Erwin Rudolf Josef Alexander Schrodinger

Schrodinger (Fig. A3.4) was born in 1887 in Vienna, Austria. His father, Rudolf Schrodinger, managed a small linoleum factory, published a paper in *The Plants and Animals Society*, and acted as vice-president of the society. Schrodinger was a cultured man and a great. Because his mother's grandfather and mother were English, English and German were used at home. Because he was educated by a private teacher, it was not necessary for him to go to school till age 11. In 1898, he entered Academy Gymnasium. Here he studied Greek and Latin as a priority and was the topper throughout.

In 1906, he entered the University of Vienna and majored in physics. He received instruction from Professor Franz Exner on experimental physics. In 1908, when listening to a lecture by Friedrich Hasenohrl, who was the successor of Boltzmann, Schrodinger was impressed by the works on Boltzmann statistics. He studied the methods of mathematical physics and the mathematical method treating the eigen value problem of partial differential equation. In 1910, he submitted a thesis to the University of Vienna and got his doctoral degree.

After service in the army of one year, after graduation, he became assistant of experiment and studied what measure was, viewing directly natural phenomena. His research spread across plenty of fields, such as electricity, influence of radioactivity on electricity in the atmosphere, sound, optics, and color. In 1914, he qualified as a professor. The First World War broke out, and he was summoned to join the army for four years. He joined the artillery stronghold at the southwest front in Austria. But he read scientific papers during his stint in the army. In November 1918, after the end of the war, he came back to the first study room of physics in the University of Vienna. In 1929, he delivered a lecture of on theoretical physics in the University of Jena, in Germany.

Fig. A3.4 Edwin Rudolf Josef Alexander Schrodinger (1887–1961). (Photograph taken in 1933.)

Soon afterward, he relocated to the University of Stuttgart. In 1921, he became a professor at the University of Zurich. There he researched on atomic structure and was impressed by the paper by de Broglie.

In 1926, Schrodinger published revolutionary research work on wave mechanics and contributed to the formulation of quantum mechanics, together with Heisenberg as mentioned above. In 1927, he succeeded Planck as the head professor of theoretical physics. There he got acquainted with Einstein. Though he was Catholic, not a Jew, he wished not to stay in Germany, which was governed by the Nazis, who were oppressing the Jews. In 1933, he relocated to Britain on an invitation to join as a professor at the University of Oxford. In the same year, he was awarded the Nobel Prize in physics. From 1936 to 1938, he served in the University of Graz but was dismissed as a politically unsuitable person by the Nazis. Afterward, he relocated to Dublin and became president of the Institute for Advanced Studies. In 1956, he became special professor of theoretical physics at the University of Vienna. In 1957, he retired (Hoffmann & Sakurayama, 1990).

—◦}}◦◦◦—◦}}◦◦◦—◦}}◦◦◦—

At the beginning of 1928, Einstein suffered from a temporary physical collapse brought on by overexertion. An enlargement of the heart was diagnosed. Hence he had to stay in bed for four months. In 1929, he built a small house in Caputh, near Berlin. There he celebrated his fiftieth birthday. Several friends together presented him a sailboat. For him, sailing on the Havel was one of his fondest pleasures (Pais, 1982, p. 317).

3.14 Princeton

The Institute for Advanced Study

In 1932, Einstein received an invitation to the Institute for Advanced Study. His plan was to first stay in Berlin for seven months and then to stay in Princeton for five months (Fig. 3.36). But this plan could not be realized because it was difficult to stay in Germany due to the rise of the Nazis. After meeting Abraham Flexner, the director of the institute, three times in October of the same year, Einstein was appointed at the Institute for Advanced Study and in December, he left Germany for the United States. When he locked up his house, he said to his wife

Fig. 3.36 Photograph of Einstein at Princeton in 1935.

that they would never see Caputh again (Pais, 1982, p. 450). The reason was that the atrocities of the Nazis were already on the rise.

On January 30, 1933, Adolf Hitler became the prime minister of Germany. On March 28, Einstein sent his resignation to the Preussen Science Academy. Then Planck, who had invited Einstein to the Preussen Science Academy, said to his secretary that even though a deep political ditch separated him from Einstein, he believed that the name of Einstein will be admired as one of the most brilliant stars in the academy for centuries into the future (Isaacson, 2017, p. 406). One week before he sent his resignation to the Preussen Science Academy, the German government carried out a raid on his house at Caputh, but only a knife for cutting bread was found (Pais, 1982, p. 450).

Life Professor at the Institute for Advanced Study

While staying at Princeton, hearing the news that Hitler had become the prime minister of Germany, Einstein decided not to go back to Germany (Raine & Okabe, 1975, p. 94). In October 1933, he became Life Professor at the Institute for Advanced Study. In 1935, he bought a residence at 112 Mercer Street, Princeton, and lived the rest of his life there. On December 20, 1936, his wife passed away after suffering from a heart disease. A short time after that, Einstein sent to Born a letter in which he explained why he was nonsocial: "I live like a bear in my cave, and really feel more at home than ever before in my eventful life. This bearlike quality has been further enhanced by the death of my woman comrade, who was better with other people than I am (Isaacson, 2017, p. 442)."

In 1938, Otto Harn (Fig. 3.37) succeeded in carrying out a nuclear fission reaction in Berlin. In 1939, the Second World War began. The same year, Einstein sent a letter warning President Roosevelt that German scientists might have acquired the

Fig. 3.37 Otto Harn (1879–1968). (Photograph taken in 1944.)

capability of constructing an atomic bomb. In October 1940, Einstein got the citizenship of the United States. In 1941, working in collaboration with the United Kingdom, the United States, under President Roosevelt, proceeded with the Manhattan Project—a secret research-and-development project for the construction of the atomic bomb. In 1944, Einstein resigned from the Institute for Advanced Study. After resigning, he continued his research at home, visiting the institute often. In 1945, atomic bombs were dropped on Hiroshima and Nagasaki and the Second World War came to an end. Einstein was very sad and upset at the misuse of the atomic bombs and one day in 1948, when he met Yukawa, who was staying in the Institute for Advanced Study, he grasped

Fig. 3.38 Einstein in 1947.

Yukawa's hand tightly and in tears apologized for the murder of innocent Japanese people with the atomic bombs (Yukawa, 1976, p. 200).

Maja

Einstein's sister, Maja Winteler, and her husband, Paul, who lived at the small estate bought by Einstein for them outside of Florence, were banished by anti-Jewish laws (racial laws) passed by Benito Mussolini. Paul relocated to Geneva, and Maja came to Princeton to live with Einstein. Soon after the end of the war, she began preparing to rejoin her husband. But in 1946, she suffered a stroke and became bedridden thereafter, never meeting her husband again. Her mind remained clear, but she could no longer speak. Einstein read to her every night after dinner (Pais, 1982, p. 473). This routine continued till July 1951, when Maja passed away.

Because of Einstein's effort to raise funds to create the Hebrew University of Jerusalem, he served as the director of the university from 1925 to 1928. In 1948, his doctor diagnosed him with abdominal aortic aneurysm. A year later, the doctor found out that the aneurysm in the abdominal aorta had become larger. In 1950, Einstein began to write his will and entrusted his scientific documents to the Hebrew University. In 1952, when the first president of Israel passed away, the government requested him to become the next president, but he rejected the request.

Einstein said that the first atomic bomb destroyed not only Hiroshima but also many people and that mankind could only be saved if a supranational system was created to eliminate the methods of brute force. One week before he passed away, he wrote a letter to Bertrand Arthur William Russell. It was a letter agreeing to sign a declaration that insisted that every nation should get rid of its nuclear weapons—a desire arising from

Einstein's wish for international peace. Now this declaration is called the Russell–Einstein Manifesto.

On April 13, 1955, Einstein collapsed when his abdominal aortic aneurysm ruptured, but he refused to be operated. On April 15, he was carried to Princeton Hospital. In the evening, a telephone call was made to his elder son, Hans Albert. Hans hurried to Princeton and arrived at the hospital the next afternoon. Since 1938, Hans had lived at Berkeley, California, and since 1947, he had served as the professor of water mechanics at the University of California, Berkeley. At a quarter past one in the morning on April 18, 1955, Einstein passed away. The funeral was performed by 12 persons, all close friends. After he was cremated, his son and friends scattered his ashes at a secret place.

Though Einstein graduated university with superior grades, he could not get a post anywhere and had a difficult time at the beginning of his research life. By the good offices of his friend Grossmann, he got a post at the Patent Office at Bern. From then on, he continued his research at home utilizing his free time after duty. In 1905, his research results were published in the form of three important papers. These publications revolutionized physics. The accumulation of his research at a place different from the academic place, like the university, where he would have liked to have got a position, made an intense impact on the academic society. He taught us that even in adversity, it is important to remain eager and endeavor to achieve your objective. His relativity theory is considered the most beautiful theory by mankind.

References

Hawking, S., & Sato, K., 佐藤勝彦. (trans. 2001). ホーキング, 未来を語る. Artist House, アーティストハウス. (2001). *The universe in a nutshell*. The Book Laboratory Inc.

Hoffmann, D., & Sakurayama, Y., 櫻山義夫. (trans. 1990). シュレーディンガーの生涯. Tijinshokan, 地人書館. (1984). Erwin Schrodinger. Verlagsgesellschaft. Leipzig.

Isaacson, W. (2017), *Einstein: His life and universe*. London: Simon & Shuster UK Ltd.

Michelson, A., & Morley, E. (1887). On the relative motion of the Earth and the luminiferous ether. *American Journal of Science*, **34**(203), 333–345.

Møller, C., & Nagata, T., & Ito, D., 永田恒夫 伊藤大介. (trans. 1959), 相対性理論. Misuzu Shobo, みすず書房. (1952). *The theory of relativity*. Oxford University Press.

Nambu, Y., 南部陽一郎. (1998). *Kuark*. 2nd ed., クォーク 第2班. Koudansha, 講談社.

Pais, A. (1982). *Subtle is the lord: The science and the life of Albert Einstein*. Oxford: Oxford University Press.

Raine, D. J., & Okabe, T., 岡部哲治. (trans.), アインシュタインと相対性理論. Tamagawa University press, 玉川大学出版部. (1975). Albert Einstein and Relativity. Wayland Publishers Ltd.

Shioyama, T., 塩山忠義. (2002). *Principle and application of sensor*, センサの原理と応用. Morikita Shuppan, 森北出版.

Tomonaga, S., 朝永振一郎. (1952). *Quantum mechanics* (Vol. 1), 量子力学 I. Misuzu Shobou, みすず書房.

Yukawa, H. (2011), 湯川秀樹. Tabibito[The Traveler], 旅人. Kadokawa Gakugei Shuppan, 角川学芸出版.

Yukawa, H., & Tamura, S. (1955-1962), 湯川秀樹 田村松平. Accepted theory of physics (Vols. I-III), 物理学通論, 上・中・下. Tokyo: Taimeidou, 大明堂.

Yukawa, S., 湯川スミ, (1976). *Garden of hardship*, 苦楽の園. Kodansha, 講談社.

Zee, A. (2010). *Quantum field theory in a nutshell*. Princeton University Press, Princeton.

Chronology of Events

Newton

1643 On Jan. 4, Isaac Newton was born in Woolsthorpe.

1646 His mother, Hannah married Barnabas Smith—this was her second marriage.

1649 Charles I was beheaded. The Puritan Revolution started.

1653 Cromwell became Lord Protector.

1653 Newton's stepfather, Smith, passed away.

1654 Newton entered King's School.

1658–1660 He took a leave of absence from King's School.

1660 Charles II took over the throne. Restoration occurred.

1661 In June, Newton entered Trinity College at University of Cambridge.

1663 Isaac Barrow joined as the first Lucasian Professor of Mathematics.

1665 Newton got his bachelor's degree in arts.

1665–1667 The university was closed due to the plague. Newton carried on research at his home in Woolsthorpe.

1667 He became a Minor Fellow.

1668 He got his master's degree in arts.

1669 He joined as a Lucasian Professor of Mathematics.

1671 He presented his own reflecting telescope to the Royal Society.

1672 His first paper, "Theory of Light and Colors," was published in the *Philosophical Transactions of the Royal Society*.

1677 Isaac Barrow passed away.

1679   Newton's mother, Hannah, passed away.

1684   Edmond Halley visited Cambridge and requested Newton for the manuscript of *De Motu Corporum in Gyrum*.

1685   James II took over the throne.

1687   *Principia* was published.

1689   William III of Orange took over the throne.

1696   Newton joined as Warden of the Royal Mint.

1699   Newton was promoted to the post of Master of the Royal Mint.

1701   He resigned the professorship of University of Cambridge.

1703   He joined as president of the Royal Society.

1705   He was knighted.

1727   On Mar. 20, he passed away in Kensington and was buried in Westminster Abbey.

Faraday

1791   On Sep. 22, Michael Faraday was born in Butts, at the edge of London, UK.

1793   King Louis XVI, of France, and his Austrian queen consort Marie-Antoinette were executed.

1800   Alessandro Volta invented the battery.

1804   Faraday was employed as a newspaper-cumerrand boy at a bookshop and stationers. Napoleon became the emperor of France.

1805   Faraday became an apprentice at the bookshop.

1810   He attended a lecture by John Tatum.

1812   He attended a lecture by Humphry Davy, completed his apprentice-ship, and relocated to DeLa Roche's shop.

1813   In Jan., he met Humphry Davy.
       On Mar. 1, he served as an experimental assistant at the Royal Institution.
       In Oct., he departed on a tour of the Continent with Humphry Davy.

1815   In Apr., Faraday ended his journey of the Continent and arrived at London.

1820   Oersted made his discovery.

1821   Faraday had success in the reverse problem of Oersted. Faraday married Salah.

1823   He succeeded in liquefying chlorine gas.

1824   He was elected as Fellow of the Royal Society.

1825   He became head of the laboratory in the Royal Institution. He discovered benzene. William Sturgeon invented the electromagnet.

1826   Faraday started giving the Christmas lectures.

1829   Humphry Davy passed away in Geneva.

1829   Faraday joined as the professor of chemistry at the English Military Academy.

1831   He submitted his paper on electromagnetic induction. It was published in 1832.

1833   He discovered the laws of electrochemical decomposition.

1836   He joined as an adviser at Trinity House.

1840   He invented a new type of chimney for oil lamps.

1844   Prime Minister Peel asked Faraday to investigate an explosion that occurred in Haswell Colliery.

1845   Faraday discovered the Faraday effect and diamagnetism.

1853   The Crimean War began. It ended in ~1856.

1854   Faraday opposed the use of chemical weapons.

1858   He was offered the use of a grace-and-favor house at Hampton Court by Queen Victoria.

1867   On Aug. 25, he passed away at Hampton Court and was buried at the High-Gate Cemetery.

Einstein

1879   On Mar. 14, Albert Einstein was born in Ulm, Germany.

1888   In Oct., he entered Luitpold-Gymnasium.

1895   He failed the entrance examination of ETH.

1896   In Oct, he entered ETH.

1900	Einstein qualified as a *fachlehrer*. Max Planck proposed the idea of an energy quantum.
1902	Einstein was employed temporally in the Patent Office in Bern. In 1904, he joined as a regular staff.
1903	He married Mileva.
1904	Hendrik Lorentz derived the Lorentz transformation.
1905	Einstein published papers on the photoelectric effect, the special relativistic theory, and Brownian motion.
1906	Ludwig Boltzmann passed away.
1907	Einstein solved a problem on the specific heat of solids.
1908	He got the post of a private lecturer.
1909	He got the post of an associate professor at Zurich University.
1911	He joined as a professor at the Karl-Ferdinand University in Prague.
1912	He joined as a professor at ETH, Zurich.
1913	He published his paper on the general relativistic theory with Marcel Grossmann. Niels Bohr proposed the atomic structure model.
1914	Einstein joined as a professor at the Berlin University. The First World War began.
1916	He published his complete version of the general relativistic theory.
1919	Arthur Eddington verified Einstein's prediction on the phenomenon of bending of light due to gravitation of the Sun by observing a total solar eclipse. Einstein divorced Mileva and married Elsa.
1921	Einstein visited the United States to raise funds to create the Hebrew University.
1922	He was awarded the Nobel Prize in Physics.
1924	De Broglie proposed the wavelike property of a particle and de Broglie's relation between wavelength and momentum.
1925	Werner Heisenberg formulated matrix mechanics.
1926	Erwin Schrodinger formulated wave mechanics.
1932	John Cockcroft and Ernest Walton at the Cavendish Institute verified mass-energy equivalence with the help of a nuclear fission reaction.

1933   Hitler became the prime minister of Germany.
Einstein became Life Professor at the Institute for Advanced Study.

1936   Elsa passed away.

1938   Otto Harn succeeded in carrying out a nuclear fission reaction.

1939   The Second World War began. Einstein sent President Roosevelt a letter notifying him of the possibility of constructing an atomic bomb.

1944   Einstein resigned from the Institute for Advanced Study.

1945   Atomic bombs were dropped in Japan. The Second World War ended.

1952   Einstein rejected the request to be the second president of Israel.

1955   On Apr. 18, he passed away in Princeton. After he was cremated, his ashes were scattered at a secret place by his son and friends.
The Russell-Einstein declaration was published.

Name Index

Subject Index

About the Author

Tadayoshi Shioyama 塩山忠義

The author got his bachelor's degree in physics from Kyoto University in 1966. In 1984, he got his doctor's degree in engineering from the Department of Mathematics and Applied Physics in Kyoto University. At present, he is professor emeritus at the Kyoto Institute of Technology.

Books published (in Japanese)

- Shioyama, T. (2002). *Principle and application of sensor,* センサの原理と応用. Morikita Shuppan, 森北出版.
- Shioyama, T. (2010). *Basic and application of image understanding and pattern recognition,* 画像理解・パターン認識の基礎と応用. Trikepus, トリケップス.
- Shioyama, T. (2019). *Newton・Faraday・Einstein—Learning physics from the great scientists' lives,* ニュートン・ファラデー・アインシュタイン—偉大な科学者の生涯から物理学を学ぶ—. Nakanishiya Shuppan, ナカニシヤ出版.

www.ingramcontent.com/pod-product-compliance
Lightning Source LLC
Chambersburg PA
CBHW061252220326
41599CB00028B/5618